Probability Theory

A Complete One-Semester Course

Probability Theory

A Complete One-Semester Course

Nikolai DOKUCHAEV
Curtin University, Australia

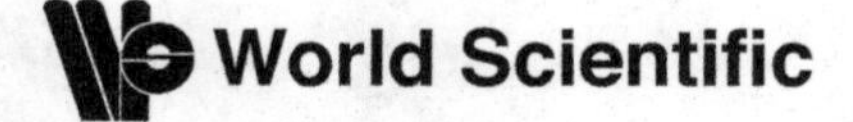

World Scientific

NEW JERSEY • LONDON • SINGAPORE • BEIJING • SHANGHAI • HONG KONG • TAIPEI • CHENNAI

Published by

World Scientific Publishing Co. Pte. Ltd.
5 Toh Tuck Link, Singapore 596224
USA office: 27 Warren Street, Suite 401-402, Hackensack, NJ 07601
UK office: 57 Shelton Street, Covent Garden, London WC2H 9HE

Library of Congress Cataloging-in-Publication Data
Dokuchaev, Nikolai.
Probability theory : a complete one-semester course / by Nikolai Dokuchaev (Curtin University, Australia).
pages cm
Includes bibliographical references and index.
ISBN 978-9814678025 (hardcover : alk. paper)
1. Probabilities--Study and teaching (Higher) 2. Mathematics--Study and teaching (Higher) I. Title.
QA273.2.D64 2015
519.2--dc23
2015013007

British Library Cataloguing-in-Publication Data
A catalogue record for this book is available from the British Library.

Printed in Singapore

Preface

This book gives a systematic, self-sufficient and yet short presentation of the mainstream topics of an introductory Probability Theory with some selected topics from Mathematical Statistics. It is suitable for a 10- to 14-week course for second- or third-year undergraduate students in Science, Mathematics, Statistics, Finance, or Economics who have completed some introductory course in Calculus, including theory of integration.

The size of the book was selected based on the following reasons. From the students' perspective, a 900-page textbook is usually very heavyweight and quite expensive; it is used by a student for one semester only, and only 20% of the text is ever read. Therefore, it is economically and environmentally sound to offer students a text that is less expensive and requires less paper to print, and that can be used on 100%. Our textbook achieves this purpose. From the lecturers' perspective, the main problem with using typical comprehensive modern textbooks is that they contain too much material. If a lecturer decides to accept a 900-page textbook in Mathematics as a course text, she will not be able to cover all material; similarly, the students will not have time to read the entire book. The lecturer will be in the position to select about 20% of the content to be covered and to be included into the course outline; this has to be done before the start of the semester. However, she cannot select arbitrarily the part to be covered; she has to study first all 900 pages. Otherwise, the selected material may rely on the materials from skipped parts. This textbook is created with the purpose to tackle this problem for lecturers; the entire textbook can be actually covered during a 10 to 14-week course.

The book contains sufficient reference material, including basic formulas and theorems, problems for tutorials and for the final exams with solutions. All necessary statistical tables are also included. To keep the course compact, we concentrate on the ultimately important core topics. In addition, to make the ideas more visible, the most technical details of the proofs are usually omitted. Moreover, a reader who wish to obtain some basic understanding of the subject may skip proofs and use the mathematical theorems as reference.

The book grew out from the *Mathematical Statistics*, *Statistical Data Analysis*, and *Statistical Inference* study units that the author taught at Curtin University. This book can be considered as a sufficient background book for the text [2]. Further reading, additional problems for tutorials and examples can be found in more comprehensive large books [3]–[7]. More detailed tables can be found in [1].

Nikolai Dokuchaev

Acknowledgments

I wish to thank all my colleagues from the Department of Mathematics and Statistics at Curtin University for their support, help, and advice. I would like to thank all my students for accepting this difficult material with understanding and patience, and for helping to improve it eventually via feedback. I would like to thank Mr. Chuong Luong for his help with the tables and pictures. Finally, I wish to thank my family, Lidia, Mikhail and Natalia, for their support.

Contents

Week 1. Probability

In this chapter, we introduce random events, probability of events, and conditional probability. In addition, we consider examples where probability can be calculated using combinatorial methods.

1.1 Probability axioms

The probability theory gives numerical values for possible scenarios under uncertainty. For a particular problem, one has to construct a probability model. For this, one has to select a sample space from a system of events and presume that each event can be assigned with a certain probability for which it can occur.

The sample space

The set of all outcomes in a probability model is called the sample space. It is usually denoted by Ω, and an element of Ω are usually denoted by ω.

Example 1.1 Consider the coin tossing game where a coin is thrown in the air and the outcome of either a head or tail is observed. The sample space for this model is the set of all possible outcomes

$$\Omega = \{h, t\}.$$

Example 1.2 A coin is thrown twice, and a sequence of heads and tails is observed. The sample space is the set of all possible outcomes

$$\Omega = \{hh, ht, th, tt\}.$$

It can be noted that a sample choice of Ω is not uniquely defined for a particular problem. Usually, it is convenient to consider the smallest possible set. However, it is also acceptable to consider larger sets Ω, including some "impossible" outcomes. This could be done for convenience or just to shorten the descriptions.

Example 1.3 Tomorrow's exchange rate for AUD/USD can be regarded as an outcome. In this case, we could select the set

$$\Omega = \{\text{rational numbers} \quad x \in [0.5, 2]\}.$$

For this example, we could also take $\Omega = \mathbf{R}$, including some "impossible" outcomes.

Definition 1.4 *Sets of outcomes (i.e. subsets of Ω) are called events (or random events).*

The standard operations of the set theory can be applied directly into probability theory.

- Let A and B be two events. The union of these two events is the event C that either A occurs or B occurs (or both occur). In terms of the set theory, $C = A \cup B$. It can also be written as "$C = A$ or B", or "$C = A + B$".
- The intersection of two events, $C = A \cap B$, is the event that both A and B occur. It can also be written as "$C = A$ and B", or "$C = A \cdot B$".
- The complement of an event, A^c, is the event that A does not occur, or $A^c = \Omega \backslash A$.
- $A \subset B$ means that A implies B (i.e., if A occurs then B occurs).
- The empty set (denoted by $\emptyset$) is the set with no elements: it is an event that does not include any outcomes. If $A \cap B = \emptyset$ (i.e., these events cannot occur at the same time), A and B are said to be disjoint.

Example 1.5 A coin is tossed twice, and a sequence of heads and tails is observed. If C is the event that there is at least one head, then $C = A \cup B$, where $A = \{th, ht\}$ and $B = \{hh\}$.

Example 1.6 If A is an event that tossing two coins have the same outcome, and B is an event that the first toss produces a head, then $A \cap B = \{hh\}$.

Example 1.7 If $A = \{hh, ht\}$, $B = \{tt, th\}$, then $C = A \cap B = \emptyset$.

Example 1.8 Let $a_i \in \mathbf{R}$ be some numbers. Then $A = \{a_1, a_2, a_3\}$ is a set, but (a_1, a_2, a_3) is an ordered set, or a three-dimensional vector. We write that $a_1 \in A$ (i.e., it is an element of A), but $\{a_1\} \subset A$ (i.e., it is a subset of A consisting of one element only).

Venn diagrams are often used for visualizing the set operations.

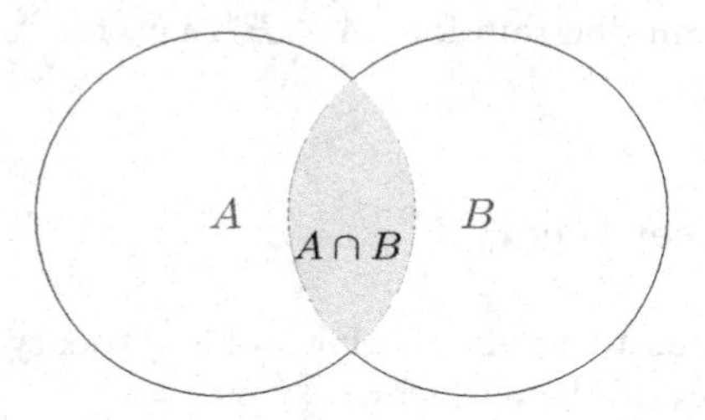

Figure 1.1: Venn diagram for "$A \cap B$", i.e., for "A and B".

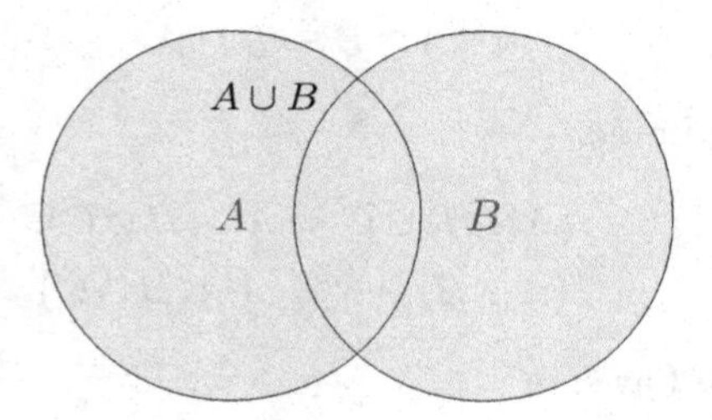

Figure 1.2: Venn diagram for "$A \cup B$", i.e., for "A or B".

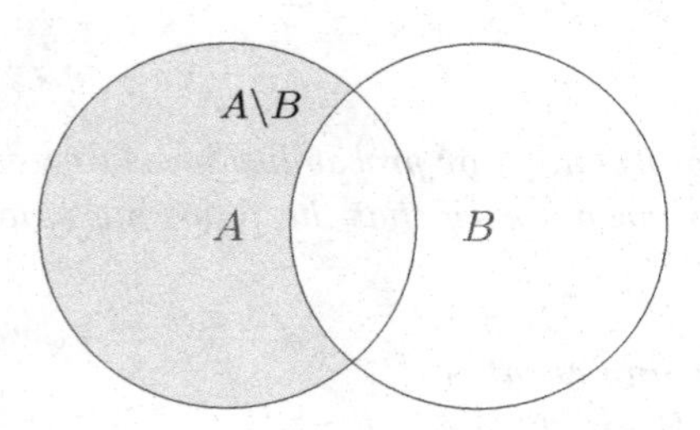

Figure 1.3: Venn diagram for "$A \backslash B$", i.e., for "A and [not B]".

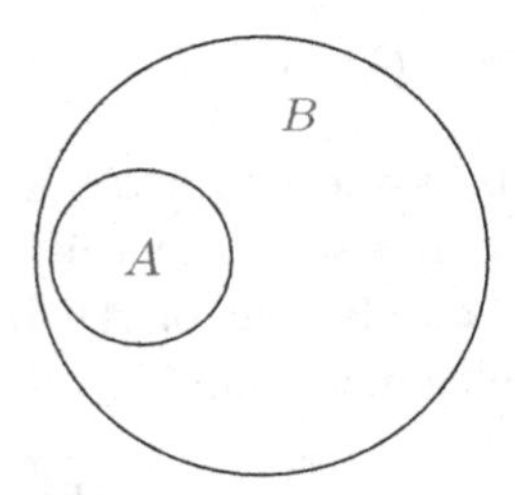

Figure 1.4: Venn diagram for "$A \subset B$", i.e., for "A implies B".

1.2 The laws of set theory

This setting allows us to apply the laws of set theory to various events. Some important laws are listed below:

◇ Commutative Laws:

$$A \cup B = B \cup A,$$
$$A \cap B = B \cap A.$$

◇ Associative Laws:

$$(A \cup B) \cup C = A \cup (B \cup C),$$
$$(A \cap B) \cap C = A \cap (B \cap C).$$

◇ Distributive Laws:

$$(A \cup B) \cap C = (A \cap C) \cup (B \cap C),$$
$$(A \cap B) \cup C = (A \cup C) \cap (B \cup C).$$

Probability

Definition 1.9 *A probability (or probability measure) is a function* $\mathbf{P}$ *that is defined on a set of events such that the following axioms are satisfied:*

(a) $\mathbf{P}(\Omega) = 1$.
(b) $\mathbf{P}(A) \in [0, 1]$ *for any event* A.
(c) *If events* A *and* B *are disjoint, then*

$$\mathbf{P}(A \cup B) = \mathbf{P}(A) + \mathbf{P}(B).$$

More generally, if A_i are mutually disjoint, $i = 1, 2, ..., \infty$, then

$$\mathbf{P}(\cup_{i=1}^{\infty} A_i) = \sum_{i=1}^{\infty} \mathbf{P}(A_i). \tag{1.1}$$

Theorem 1.10 *The axioms of probability imply the following properties.*
(A) $\mathbf{P}(A^c) = 1 - \mathbf{P}(A)$.
(B) $\mathbf{P}(\emptyset) = 0$.
(C) If $A \subset B$ *then* $\mathbf{P}(A) \leq \mathbf{P}(B)$.
(D) Addition Law: $\mathbf{P}(A \cup B) = \mathbf{P}(A) + \mathbf{P}(B) - \mathbf{P}(A \cap B)$.

Proof. Property A follows from the axiom and the fact that

$$\Omega = A \cup A^c, \quad \mathbf{P}(\Omega) = 1 = \mathbf{P}(A) + \mathbf{P}(A^c).$$

Property B follows from Property A since $\emptyset = \Omega^c$.

Properties C-D can be seen from the Venn diagrams 1.2 and 1.4 respectively.

Remark 1.11 *It has to be clarified, that, in many important models, there are subsets of* Ω *that are not considered to be random events; these subsets are such that the probability for them is not defined. The choices of the systems of events are defined by the information flow in the model. More advanced theory of stochastic analysis discusses these choices in detail.*

Example of computing probabilities: Counting method

Let us consider the simplest case where

$$\Omega = \{\omega_1, \omega_2, ..., \omega_N\}$$

and

$$\mathbf{P}(\{\omega_k\}) = 1/N.$$

In this case, we compute the probabilities by the so-called *counting method.* To find the probability of an event $A = \{\omega\}$ we simply add the probabilities of the ω_k that constitute A.

In other words, if A has n elements (i.e., A can occur in any of n mutually exclusive ways), then

$$\mathbf{P}(A) = \frac{n}{N}.$$

Note that this formula holds only if all the outcomes are equally likely (i.e. having the same probability of occurring).

Example 1.12 Suppose that a fair coin is thrown twice and the sequence of heads and tails is recorded. The sample space is

$$\Omega = \{hh, ht, th, tt\}.$$

We assume that each elementary outcome in Ω has probability $1/4$ (i.e., they are equally likely). Let $A = \{ht, hh\}$ be the event of getting heads on the first toss. Then $\mathbf{P}(A) = 2/4 = 1/2$. Let $B = \{hh, th\}$ be the event of getting heads on the second toss. Then $\mathbf{P}(B) = 2/4 = 1/2$.

Example 1.13 In the previous example, let $C = A \cup B$ be the event that heads comes up on the first toss or on the second toss. Clearly,

$$\mathbf{P}(C) = \mathbf{P}(A) + \mathbf{P}(B) - \mathbf{P}(A \cap B) = 1/2 + 1/2 - 1/4 = 3/4.$$

Let D denote the event that at least one head is thrown. Then $D = \{hh, ht, th\}$ and $\mathbf{P}(D) = 0.75$.

Example 1.14 In the previous example, if one only records the number of heads, then Ω would be $\{0, 1, 2\}$. These outcomes are not equally likely, and $\mathbf{P}(A) \neq 2/3$.

It can be seen that it is important to be able to count the number of outcomes. It requires to use some combinatorial rules and formulas.

1.3 Conditional probability

Definition 1.15 *Let A and B be two events with $\mathbf{P}(B) > 0$. The conditional probability of A given B is defined to be*

$$\mathbf{P}(A|B) = \frac{\mathbf{P}(A \cap B)}{\mathbf{P}(B)}.$$

The idea behind this definition is that if the event occurs, the relevant sample space becomes B rather than Ω and the conditional probability is a probability measure on B.

Theorem 1.16 (Multiplication law) *Let A and B be two events with $\mathbf{P}(B) > 0$. Then*

$$\mathbf{P}(A \cap B) = \mathbf{P}(A|B)\mathbf{P}(B).$$

In some situations, $\mathbf{P}(A|B)$ and $\mathbf{P}(B)$ can be found rather easily. We can then find $\mathbf{P}(A \cap B)$ by the multiplication law.

Example 1.17 An urn contains 3 red balls and 1 blue ball. Two balls are selected without replacement. Let us find the probability that they are both red. Let R_1 and R_2 denote the events that a red ball is drawn on the first trial and on the second trial, respectively. From the multiplication law,

$$\mathbf{P}(R_1 \cap R_2) = \mathbf{P}(R_1)\mathbf{P}(R_2|R_1).$$

Clearly, $\mathbf{P}(R_1) = 3/4$, and if a red ball has been removed on the first trial, there are 2 red balls and 1 blue ball left. Therefore $\mathbf{P}(R_2|R_1) = 2/3$ and $\mathbf{P}(R_1 \cap R_2) = 1/2$.

1.4 Bayes' rules

Bayes' rule I (the Law of Total Probability)

Theorem 1.18 *Let events B_1, B_2,...,B_n be such that $\cup_i B_i = \Omega$ and $B_i \cap B_j = \emptyset$ for $i \neq j$ (disjoint events), with $\mathbf{P}(B_j) > 0$ for all j. Then, for any event A,*

$$\mathbf{P}(A) = \mathbf{P}(A|B_1)\mathbf{P}(B_1) + \mathbf{P}(A|B_2)\mathbf{P}(B_2) + \cdots + \mathbf{P}(A|B_n)\mathbf{P}(B_n).$$

Proof.

$$\begin{aligned}\mathbf{P}(A) &= \mathbf{P}(A \cap \Omega) = \mathbf{P}(A \cap (\cup_i B_i)) = \mathbf{P}(\cup_i (A \cap B_i)) \\ &= \sum_i \mathbf{P}(A \cap B_i) = \sum_i \mathbf{P}(B_i)\mathbf{P}(A|B_i).\end{aligned}$$

We use here that the events $A \cap B_i$ are disjoint for different i.

This theorem is useful in situations where it is difficult to calculate $\mathbf{P}(A)$ directly but $\mathbf{P}(A|B_i)$ and $\mathbf{P}(B_i)$ are easy to find.

Example 1.19 An urn contains 3 red balls and 1 blue ball. Two balls are selected without replacement, as in the example above. Let us find the probability that a red ball is selected on the second draw?

By the Law of Total Probability,

$$\mathbf{P}(R_2) = \mathbf{P}(R_2|R_1)\mathbf{P}(R_1) + \mathbf{P}(R_2|R_1^c)\mathbf{P}(R_1^c) = \frac{2}{3} \times \frac{3}{4} + 1 \times \frac{1}{4} = \frac{3}{4}.$$

Here R_1^c is the compliment of R_1, i.e., it is the event that a blue ball is drawn on the first trial.

Bayes' rule II

Theorem 1.20 *Let A and B be events, with $\mathbf{P}(A) > 0$ for all j. Then*

$$\mathbf{P}(B|A) = \frac{\mathbf{P}(A|B)\mathbf{P}(B)}{\mathbf{P}(A)}.$$

Proof. Observe that $\mathbf{P}(A \cap B) = \mathbf{P}(B|A)\mathbf{P}(A) = \mathbf{P}(A|B)\mathbf{P}(B)$.

Bayes' rule III

Theorem 1.21 *Let A and B_1,..., B_n be events where the B_j are disjoint, $\cup_i B_i = \Omega$, with $\mathbf{P}(A) > 0$ for all j. Then*

$$\mathbf{P}(B_j|A) = \frac{\mathbf{P}(A|B_j)\mathbf{P}(B_j)}{\sum_i \mathbf{P}(B_i)\mathbf{P}(A|B_i)}.$$

Proof. Apply the previous rule to $B = B_j$ and observe that

$$\mathbf{P}(A) = \sum_i \mathbf{P}(B_i)\mathbf{P}(A|B_i).$$

Example 1.22 An urn contains 3 red balls and 2 blue balls. Two balls are selected one after another without replacement. It is known that the 2nd selected ball is red. What is the probability that the first ball is also red?

Again, let R_1 and R_2 denote the events that a red ball is drawn on the first trial and on the second trial, respectively. From the multiplication law,

$$\mathbf{P}(R_1 \cap R_2) = \mathbf{P}(R_1)\mathbf{P}(R_2|R_1).$$

By Bayes' rule, we have that

$$\mathbf{P}(R_1|R_2) = \frac{\mathbf{P}(R_2|R_1)\mathbf{P}(R_1)}{\mathbf{P}(R_2)} = \frac{\mathbf{P}(R_2|R_1)\mathbf{P}(R_1)}{\mathbf{P}(R_1)\mathbf{P}(R_2|R_1) + \mathbf{P}(R_1^c)\mathbf{P}(R_2|R_1^c)}.$$

Clearly, $\mathbf{P}(R_1) = 3/5$, $\mathbf{P}(R_1^c) = 2/5$. If a red ball has been removed on the first trial, there are 2 red balls and 2 blue balls left. Therefore, $\mathbf{P}(R_2|R_1) = 2/4 = 1/2$. If a blue ball has been removed on the first trial, there are 3 red balls and 1 blue ball left. Therefore, $\mathbf{P}(R_2|R_1^c) = 3/4$, and

$$\mathbf{P}(R_1|R_2) = \frac{1/2 \times 3/5}{3/5 \times 1/2 + 2/5 \times 3/4} = \frac{1}{2}.$$

1.5 Independence

Definition 1.23 *A and B are said to be independent events if*

$$\mathbf{P}(A \cap B) = \mathbf{P}(A)\mathbf{P}(B).$$

Intuitively, we would say that two events, A and B, are independent if knowing that one had occurred gave us no information about whether the other had or had not occurred; that is,

$$\mathbf{P}(A|B) = \mathbf{P}(A).$$

This is equivalent to the definition of independency (if $\mathbf{P}(B) \neq 0$).

Example 1.24 A card is selected randomly from a deck. Let

$$A = \{\text{it is an ace}\}, \qquad D = \{\text{it is a diamond}\}.$$

We have $\mathbf{P}(A) = 1/13$ and $\mathbf{P}(D) = 1/4$. Now, let $B = A \cap D$ be the event that the card is the ace of diamonds where $\mathbf{P}(B) = 1/52$. We have

$$\mathbf{P}(A)\mathbf{P}(D) = \mathbf{P}(A \cap D) = 1/52.$$

Hence the events are A and D independent.

Example 1.25 A system is designed so that it fails only if a unit and a backup unit both fail. Assuming that these failures are independent and that each unit fails with probability p, the system fails with probability p^2. If, for example, the probability that any unit fails during a given year is .1, then the probability that the system fails is .01, which represents a considerable improvement in reliability.

Definition 1.26 *We define a collection of events, $A_1, A_2, ..., A_n$ to be mutually independent if, for any sub-collection,*

$$\mathbf{P}(A_{m_1} \cap A_{m_2} ... \cap A_{m_k}) = \mathbf{P}(A_{m_1})\mathbf{P}(A_{m_2}) \cdots \mathbf{P}(A_{m_k}).$$

1.6 Combinatorial formulas

The multiplication principle

Theorem 1.27 *If one experiment has m outcomes and another experiment has n outcomes, then there are $m \times n$ possible outcomes for the two experiments.*

Proof. Denote the outcomes of the first experiment by $a_1, ..., a_m$ and the outcomes of the second experiment by $b_1, ..., b_n$. The outcomes for the two experiments are the ordered pairs (a_i, b_j). These pairs can be exhibited as the entries of an $m \times n$ rectangular array, in which the pair (a_i, b_j) is in the ith row and the jth column. Hence, there are mn entries in this array.

Example 1.28 A class has 12 boys and 18 girls. The teacher selects 1 boy and 1 girl to act as representatives to the student government. She can do this in any of $12 \times 18 = 216$ different ways.

Extended multiplication principle

Theorem 1.29 *If there are p experiments and the first has n_1 possible outcomes, the second n_2, and the pth has n_p possible outcomes, then there is a total of $n_1 n_2 \cdots n_p$ possible outcomes for the p experiments.*

Proof. We will use the induction method. The statement is true for $p = 2$. Assume that it is true for $p = q$, i.e., that there are $n_1 n_2 \cdots n_q$ possible outcomes for the first q experiments.

To complete the proof by induction, we must show that it follows that the property holds for $p = q+1$. Apply the multiplication principle, regarding the first q experiments as a single experiment with $n_1 n_2 \cdots n_q$ outcomes. It follows that there are $(n_1 n_2 \cdots n_q) n_{q+1}$ outcomes for the $q + 1$ experiments.

Example 1.30 Consider a sequence of 6 digits, of which each may be either a 0 or a 1 (we may call it a 6-bit binary word). How many different 6-bit words are there?

There are two choices for the first bit, two for the second, etc., and thus there are $2 \times 2 \times \cdots \times 2 = 2^6 = 64$ such words.

Permutations and combinations

Definition 1.31 *A permutation is an ordered arrangement of objects.*

Definition 1.32 *(a) Sampling without replacement is sampling such that no duplication is allowed.*

(b) Sampling with replacement is sampling such that duplication is allowed.

Example 1.33 Consider taking balls from an urn. Without replacement, we are not allowed to put a ball back before choosing the next one, but with replacement, we can do so.

The extended multiplication principle can be used to count the number of different ordered samples possible for a set of n elements.

Theorem 1.34 *(a) For a set of size n and a sample of size r, there are n^r different ordered samples with replacement.*
(b) For a set of size n and a sample of size r, there are

$$n(n-l)(n-2)\cdots(n-r+1)$$

different ordered samples without replacement.

Proof. First, suppose that sampling is done with replacement. The first ball can be chosen in any of n ways, the second in any of n ways, etc., so that there are

$$n \times n \times \cdots \times n = n^r$$

samples. Next, suppose that sampling is done without replacement. There are n choices for the first ball, $n-1$ choices for the second ball, $n-2$ for the third, and $n-r+1$ for the rth.

Corollary 1.35 *The number of orderings of n elements is*

$$n(n-1)(n-2)\cdots 1 = n!.$$

Example 1.36 Let us find the number of ways for four soldiers to be lined up. This is sampling without replacement. According to the above corollary, there are $4! = 4 \times 3 \times 2 \times 1 = 24$ different lines.

Example 1.37 Suppose that from 9 soldiers, 5 are to be chosen and lined up. In this case, there are $9 \times 8 \times 7 \times 6 = 3024$ different lines.

Example 1.38 Suppose that license plates have 6 characters: 3 letters followed by 3 numbers. How many distinct such plates are possible?

Solution. This corresponds to sampling with replacement. There are $26^3 = 17,576$ different ways to choose the letters and $10^3 = 1000$ ways to choose the numbers. By the multiplication principle, there are $17,576 \times 1000 = 17,576,000$ different plates.

Example 1.39 If, under the assumptions of the previous example, all sequences of six characters are equally likely, what is the probability that the license plate for a new car will contain no duplicate letters or numbers?

Solution. Call the desired event A; Ω consists of all 17,576,000 possible sequences. Since these are all equally likely,

$$\mathbf{P}(A) = \frac{n}{17,576,000},$$

where n is the number of ways that A can occur. There are 26 choices for the 1st letter, 25 for the 2nd, 24 for the 3rd, and hence $26 \times 25 \times 24 = 15,600$ ways to choose the letters without duplication (doing so corresponds to sampling without replacement), and $10 \times 9 \times 8 = 720$ ways to choose the numbers without duplication. From the multiplication principle, $n = 15,600 \times 720 = 11,232,000$. Thus

$$\mathbf{P}(A) = \frac{11,232,000}{17,576,000} = 0.64.$$

Unordered samples (combinations)

We considered the above counting permutations; now, we will study how to count *combinations*, where the order in which they were obtained is not considered.

Theorem 1.40 *The number of unordered samples of r objects selected from n objects without replacement is*

$$\frac{n(n-1)\cdots(n-r+1)}{r!} = \frac{n!}{(n-r)!r!}. \tag{1.2}$$

Proof. From the multiplication principle, the number of ordered samples equals the number of unordered samples multiplied by the number of ways to order each sample. The number of ordered samples is $n(n-1)\cdots(n-r+1)$, and a sample of size r can be ordered in is $r!$.

Values (1.2) are usually denoted as C_n^r as well as $\binom{n}{r}$; they are called the binomial coefficients, because they occur in the expansion

$$(a+b)^n = \sum_{r=0}^{n} \binom{n}{r} a^r b^{n-r}.$$

Example 1.41 A committee of 7 members is to be divided into two subcommittees of size 3 and 4. This can be done in

$$\frac{7!}{3!4!} = 35$$

ways.

The previous theorem can be generalized as follows.

Theorem 1.42 *The number of ways that n objects can be grouped into r classes with n_i elements in the ith class, $i = 1, ..., r$, and $\sum_i n_i = n$, is*

$$\frac{n!}{n_1! \times n_2! \cdots \times n_r!}.$$

Example 1.43 A committee of 7 members is to be divided into three subcommittees of size three, two, and two. This can be done in

$$\frac{7!}{3!2!2!} = 210$$

ways.

Problems for Week 1

Problem 1.1 *Calculate* $\mathbf{P}(A \cap B)$ *given that* $\mathbf{P}(A) = 0.5$, $\mathbf{P}(B) = 0.4$, *where A and B are independent events.*

Problem 1.2 *Calculate* $\mathbf{P}(A \cup B)$ *given that* $\mathbf{P}(A) = 0.8$, $\mathbf{P}(B) = 0.4$, *where A and B are independent events.*

Problem 1.3 *Show that if A and B are independent events then A and B^c are also independent events. Remind that $B^c = \Omega \backslash B$ (i.e., [not B]).*

Problem 1.4 *Calculate* $\mathbf{P}(A \cap B^c)$ *given that* $\mathbf{P}(A) = 0.5$, $\mathbf{P}(B) = 0.4$, *where A and B are independent events.*

Problem 1.5 *Calculate* $\mathbf{P}(A \backslash B)$ *given that* $\mathbf{P}(A) = 0.5$, $\mathbf{P}(B) = 0.4$, *where A and B are independent events.*

Problem 1.6 *A manufacturer produces parts on two machines. Machine A produces 40% of the parts while the rest are produced on machine B. 10% of the parts produced by machine A are defective while 5% of the parts produced by machine B are defective. What is the probability that a part is produced by machine B given that it is defective?*

Week 2. Random Variables

In this chapter, we introduce random variables, including discrete and continuous random variables.

2.1 Random variables and their distributions

A random variable is essentially a random number. More precisely, it is a mapping $X : \Omega \to \mathbf{R}$.

Example 2.1 A coin is thrown twice and the sequence of heads and tails is observed

$$\Omega = \{hh, ht, th, tt\}.$$

Examples of random variables defined on Ω are

- the total number of heads X;
- the total number of tails Y;
- the number of heads minus the number of tails, i.e., $Z = X - Y$;
- number 1 (since any non-random constant can be considered as trivial case of a random variable, or a constant function defined on Ω).

Each of these is a real-valued function defined on Ω; each is a rule that assigns a real number to every $\omega \in \Omega$.

Example 2.2 Let X be the total number of heads in the previous example. If the coin is fair, then each of the outcomes in Ω above has the probability 1/4, and

$$\mathbf{P}(X = 0) = \frac{1}{4}, \quad \mathbf{P}(X = 1) = \frac{1}{2}, \quad \mathbf{P}(X = 2) = \frac{1}{4}.$$

Definition 2.3 *The distribution of a random variable is the set of all probabilities* $\mathbf{P}(a < X \leq b)$ *for all* a *and* b *such that* $-\infty \leq a < b \leq +\infty$.

It must be clarified that the identity of two probability distributions does not mean identity of the random variables for which they belong. For example, X and $-X$ have the same distribution if X is *standard normal* (we will discuss these random variables below).

Definition 2.4 *The cumulative distribution function (c.d.f.) of a random variable* X *is*

$$F(x) = \mathbf{P}(X \leq x), \quad x \in \mathbf{R}.$$

We may use notations $F_X(x)$, $F_Y(x)$, $F_Y(y)$ for the c.d.f.'s of X and Y.

Example 2.5 The c.d.f. for X from Example 2.2 is

$$\begin{aligned} F(x) &= \mathbf{P}(X \leq x) = 0, \quad x < 0, \\ F(x) &= \mathbf{P}(X \leq x) = \tfrac{1}{4}, \quad x \in [0,1), \\ F(x) &= \mathbf{P}(X \leq x) = \tfrac{3}{4}, \quad x \in [1,2), \\ F(x) &= \mathbf{P}(X \leq x) = 1, \quad x \geq 2. \end{aligned}$$

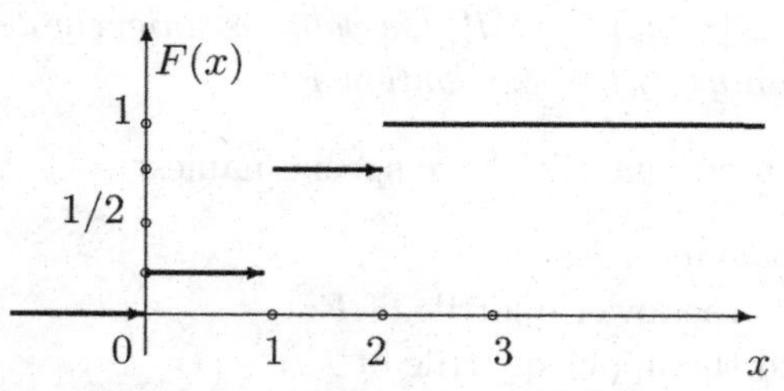

Figure 2.1: The shape of the c.d.f. $F(x)$ for Example 2.2.

Further, the cumulative distribution functions have the following properties:

- any c.d.f. $F(x)$ is non-decreasing;
- $\lim_{x\to-\infty} F(x) = 0, \qquad \lim_{x\to+\infty} F(x) = 1.$
- any c.d.f. $F(x)$ is right-continuous; i.e.,

$$\lim_{\varepsilon\to 0,\ \varepsilon>0} F(x+\varepsilon) = F(x).$$

This means that it is a function which is continuous at all points when approached from the right. This can be written as

$$F(x+0) = F(x).$$

Moreover, for any function with these properties, a random variable can be found that this function is its c.d.f.

The cumulative distribution function describes completely the probability distribution of a random variable.

2.2 Some quantitative characteristics: Median and quantiles

Definition 2.6 *Let M be a value such that $\mathbf{P}(X \leq M) = 0.5$. If M is uniquely defined, then this value is called the median of the distribution.*

If F is monotonically increasing, then the median is uniquely defined as $M = F^{-1}(0.5)$ (i.e., it is the inverse of the function F). It may happen that there exists an interval (c_1, c_2) such that $F(c_1) = F(c_2) = 0.5$. In this case, we select the largest interval of this kind and call the value $M = (c_1 + c_2)/2$ to be the median. In this sense, the median is uniquely defined.

Definition 2.7 *Let $p \in (0,1)$ be given, and let x_p be a value such that $F(x_p) = p$, or $\mathbf{P}(X \leq x_p) = p$. If this value is uniquely defined, then it is called the pth quantile of the distribution F.*

Some commonly used quantiles have special names:

- $x_{0.5}$ is the median of F;
- $x_{0.25}$ is called the lower quartile of F;
- $x_{0.75}$ is called the upper quartile of F.

2.3 Independent random variables

Definition 2.8 *Two random variables X and Y are said to be independent if, for all $a_1, b_1, a_2, b_2 \in \mathbf{R}$, the events*

$$X \in (a_1, b_1] \quad \text{and} \quad Y \in (a_2, b_2]$$

are independent, or, equivalently,

$$\mathbf{P}(X \in (a_1, b_1] \quad \text{and} \quad Y \in (a_2, b_2]) = \mathbf{P}(X \in (a_1, b_1])\,\mathbf{P}(Y \in (a_2, b_2]).$$

Theorem 2.9 *X and Y are independent if and only if*

$$\mathbf{P}(X \le x \quad \text{and} \quad Y \le y) = F_X(x)F_Y(y).$$

Theorem 2.10 *Given $g : \mathbf{R} \to \mathbf{R}$ and $h : \mathbf{R} \to \mathbf{R}$ are functions, if X and Y are independent random variables, then the random variables $g(X)$ and $h(Y)$ are independent as well.*

2.4 Discrete random variables

Definition 2.11 *A countably infinite set is a set that can be put into one-to-one correspondence with the natural numbers (i.e., can be "counted"). We say that a set is countable if it is either finite or finitely countable.*

Example 2.12 The set of all integers is countable. The set of all rational numbers is also countable. The set of the point of the real interval $(0, 1) \subset \mathbf{R}$ is not countable.

Definition 2.13 *A discrete random variable is a random variable that can take on only a finite or at most a countably infinite number of values.*

Example 2.14 Let X be the total number of tosses in an experiment that consists of tossing a coin 5 times. The possible values of X are 0, 1, 2, 3, 4, 5.

Example 2.15 Let X be the total number of tosses in an experiment that consists of tossing a coin until a head turns up. The possible values of X are $0, 1, 2, 3, 4, 5, 6, 7, \ldots$, i.e., it takes values at a countably infinite set.

Theorem 2.16 *For discrete random variables, their c.d.f. are piecewise constant (i.e., step functions).*

This can be illustrated by the diagram in Figure 2.1 for Example 2.2.

Definition 2.17 *Let a random variable X takes values $x_1, x_2, x_3, \ldots$ with probabilities $p_1, p_2, p_3, \ldots$ respectively. The function $p(x)$ on $\mathbf{R}$ such that*

$$p(x_i) = p_i$$

and

$$p(x) = 0, \quad x \notin \cup\{x_i\}$$

is called the probability mass function, *or the* probability frequency function, *of the random variable X.*

Note that the c.d.f. jumps wherever $p(x) > 0$ and that the size of the jump at x is $p(x)$.

The probability frequency function describes completely the probability distribution of the discrete random variable; it is usually more convenient than the cumulate distribution function (c.d.f.). Figure 2.2 demonstrates the diagram for the probability frequency function for Example 2.2.

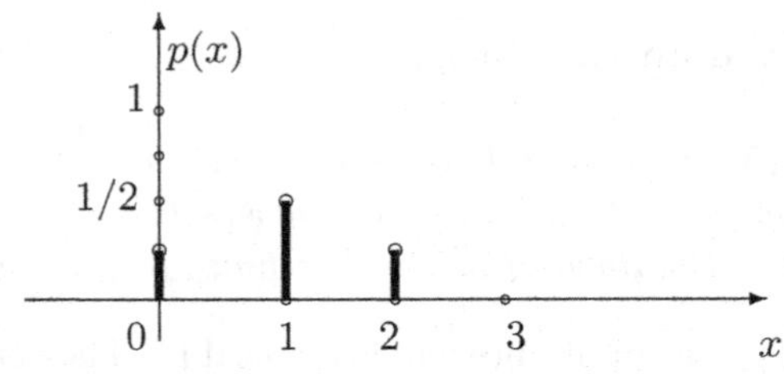

Figure 2.2: The shape of the probability frequency function $p(x)$ for Example 2.2.

Independent discrete random variables

Definition 2.18 *Consider two discrete random variables X and Y taking on possible values $x_1, x_2, ...,$ and $y_1, y_2, ...$ We say that X and Y are independent if, for all i and j,*

$$\mathbf{P}(X = x_i \text{ and } Y = y_j) = \mathbf{P}(X = x_i)\mathbf{P}(Y = y_j).$$

2.5 Examples of discrete distributions

Discrete uniform distribution

Assume that a finite set $\{a_1, a_2, ..., a_n\}$ is given. A random variable X such that $\mathbf{P}(X = a_i) = 1/n$ is said to have the uniform distribution on the set $a_1, a_2, ..., a_n$. For this distribution, all points are equally likely.

Bernoulli random variables

Definition 2.19 *A Bernoulli random variable takes on only two values: 0 and 1, with probabilities $1 - p$ and p, respectively. Its frequency function is thus*

$$p(x) = \begin{cases} p, & x = 1 \\ 1 - p, & x = 0. \end{cases}$$

Binomial distribution

Definition 2.20 *A random variable X is said to have a binomial distribution with parameters n and p, where $p \in (0,1)$, $n > 0$ is an integer, if*

$$\mathbf{P}(X = k) = p(k) = \binom{n}{k} p^k (1-p)^{n-k}.$$

In this case, X is said to be a binomial random variable; $X \sim Bin(n,p)$.

Theorem 2.21 *Suppose that n independent experiments, or trials, are performed, where n is a fixed number and that each experiment results in a "success" with probability p and a "failure" with probability $1-p$. Then the total number of successes, X, has the binomial distribution.*

Proof. Any particular sequence of k successes occurs with probability $p^k(1-p)^{n-k}$, from the multiplication principle. The total number of such sequences is $C_n^k = \binom{n}{k}$, since there are C_n^k ways to assign k successes to n trials.

Example 2.22 A coin is tossed 5 times and the total number of heads is counted ("head" is identified with "success"). Then

$$\mathbf{P}(X = 2) = \binom{5}{2} 0.5^2 (1-0.5)^{5-2} = 0.3125.$$

A random variable with a binomial distribution can be expressed in terms of independent Bernoulli random variables.

Let $X_1, ..., X_n$ be independent Bernoulli random variables with

$$\mathbf{P}(X_i = 1) = p, \quad \mathbf{P}(X_i = 0) = 1 - p.$$

Then

$$Y = X_1 + \cdots + X_n$$

is a binomial random variable.

Poisson distribution

Definition 2.23 *Let $\lambda > 0$ be given. We say that a discrete random variable X has a Poisson frequency function with parameter λ if*

$$\mathbf{P}(X = k) = \frac{\lambda^k}{k!} e^{-\lambda}, \quad k \geq 0.$$

Note that, since the Taylor series dei the exponent is $e^{\lambda} = \sum_{k=0}^{+\infty} \frac{\lambda^k}{k!}$, it follows that the frequency function sums to 1.

Theorem 2.24 *The Poisson distribution can be derived as the limit of a binomial distribution with the number of trials n and the probability of success on each trial p if there exists $\lambda > 0$ such that*

$$p \to 0, \quad n \to +\infty, \quad \lambda = pn.$$

Proof. Let the binomial frequency function be

$$p(k) = \frac{n!}{(n-k)!k!} p^k (1-p)^{n-k}.$$

It can be rewritten as

$$p(k) = \frac{n!}{(n-k)!k!} \left(\frac{\lambda}{n}\right)^k \left(1 - \frac{\lambda}{n}\right)^{n-k} = \frac{\lambda^k}{k!} \left(1 - \frac{\lambda}{n}\right)^n A_n,$$

where

$$A_n = \frac{n!}{(n-k)!n^k} \left(1 - \frac{\lambda}{n}\right)^{-k} \to 1,$$

$$\left(1 - \frac{\lambda}{n}\right)^n \to e^{-\lambda} \quad \text{as} \quad n \to +\infty.$$

Example 2.25 Two dice are rolled 100 times, and the number of double sixes, X, is counted. The distribution of X is binomial with $n = 100$ and $p = (1/6)^2 = 0.0278$. Since n is large and p is small, we can approximate the binomial probabilities by Poisson probabilities with $\lambda = np = 2.78$.

The exact binomial probabilities and the Poisson approximations can be found, for instance, in the tables [1].

Theorem 2.26 *Assume that the number X of some events occurring in a time interval has Poisson distribution with mean occurrence λ per unit time. Let Y be the number of events occurring in s units of time. Then Y has Poisson distribution $Poi(s\lambda)$.*

Proof. We have $\lambda = np$ for the given time interval of the size 1, where p is a probability that there is a call during a small time interval Δt, and $n = 1$ hour$/\Delta t$. Therefore, the distribution of X can be approximated as $Bin(n, p)$. For the time interval $s \times 1$, we have $n_s = ns$ of these small time periods. Therefore, the distribution of Y can be approximated as $Bin(ns, p)$, i.e., it is a Poisson distribution $Poi(pns)$, or $Poi(s\lambda)$, where $\lambda = pn$.

Geometric binomial distribution

The geometric distribution is constructed from infinite sequence of independent Bernoulli trials as follows.

Definition 2.27 *Let $p \in (0,1)$ be given. A random variable X is said to have a geometric binomial distribution if*

$$p(k) = \mathbf{P}(X = k) = (1-p)^{k-1}p, \quad k = 1, 2, 3, ...$$

Note that the sequence of probabilities is a geometric sequence.

Theorem 2.28 *Consider infinite sequence of independent Bernoulli trials. On each trial, a success occurs with probability p, and X is the total number of trials up to and including the first success. X will have a geometric binomial distribution.*

Proof. If $X = k$, then there are $k-1$ failures followed by a success. From the independence of the trials, this occurs with probability

$$p(k) = \mathbf{P}(X = k) = (1-p)^{k-1}p, \quad k = 1, 2, 3, ...$$

Example 2.29 If the probability of winning in a lottery is 1/9, then the distribution of the number of tickets a person must purchase up and including the first winning ticket is a geometric with $p = 1/9$.

Negative binomial distribution

Definition 2.30 *Let $p \in (0,1)$ and an integer $r > 0$ be given. A random variable X is said to have a negative binomial distribution if*

$$\mathbf{P}(X = k) = \binom{k-1}{r-1} p^r (1-p)^{k-r}, \quad \textit{for} \quad k \geq r,$$

$$\mathbf{P}(X = k) = 0, \quad \textit{for} \quad k < r.$$

Theorem 2.31 *Consider a sequence of independent trials with probability of success p performed until there are exactly r successes. Let X be the total number of trials. Then X have a negative binomial distribution.*

Proof. Let us find $\mathbf{P}(X = k)$. Any such particular sequence has probability $p^r(1-p)^{k-r}$, from the independence assumption. The last trial is a success, and the remaining $r-1$ successes can be assigned to the remaining $k-1$ trials in $\binom{k-1}{r-1}$ ways.

Hypergeometric distribution

Definition 2.32 *Let positive integers n, m, r be given. We say that a random variable X has a hypergeometric distribution if*

$$\mathbf{P}(X = k) = \frac{\binom{r}{k}\binom{n-r}{m-k}}{\binom{n}{m}}, \quad k \leq r.$$

Example 2.33 Suppose that an urn contains n balls, of which r are black and $n - r$ are white. Let X denote the number of black balls drawn when taking m balls without replacement. Then X has a hypergeometric distribution.

2.6 Continuous distributions

In applications, we are often interested in random variables that can take on a continuum of values rather than a finite or countably infinite number.

Definition 2.34 *A random variable X is said to be continuous if $\mathbf{P}(X = x) = 0$ for any $x \in \mathbf{R}$. Respectively, a probability distribution is called continuous if it belongs to a continuous random variable.*

Remark 2.35 This definition may seem as a paradox given property (1.1) in the probability axioms, a paradox, since $\mathbf{P}(X \in \mathbf{R}) = 1$, and the event $\{X \in \mathbf{R}\}$ can be represented as the union of the disjoint events $\{X = x\}$, i.e. $\{X \in \mathbf{R}\} = \cup_{x \in \mathbf{R}}\{X = x\}$. On the other hand, $\mathbf{P}(X = x) = 0$ for all $x \in \mathbf{R}$. However, there is no contradiction here: property (1.1) was imposed for countable sets $\{A_j\}$ only, and the set $\{x\} = \mathbf{R}$ is non-countable.

In fact, X is a continuous random variable if and only if[1] its c.d.f. is a continuous function. In particular, for continuous random variables,

$$\mathbf{P}(X = x) = \mathbf{P}(X \leq x) - \mathbf{P}(X < x) = 0.$$

Definition 2.36 *A distribution is called continuous if it belongs to a continuous random variable.*

Definition 2.37 *A distribution and the corresponding random variable is called absolutely continuous if*

$$F(x) = \int_{-\infty}^{x} f(y)dy$$

[1] Sometimes, "if and only if" is shortened as "iff".

for some integrable function f.

Definition 2.38 *The derivative $f(x) = dF(x)/dx$ is said to be a density function (or probability density function, or p.d.f.).*

For the density, we have that

$$\mathbf{P}(a < X < b) = \mathbf{P}(a \le X < b) = \mathbf{P}(a < X \le b) = \mathbf{P}(a \le X \le b)$$
$$= \int_a^b f(x)dx = F(b) - F(a),$$

for any $-\infty \le a \le b \le +\infty$. Therefore, the distribution of X can be described by its p.d.f.

For an absolutely continuous random variable, the role of the frequency function is taken by $f(x)$, which has the properties that

$$f(x) \ge 0, \qquad \int_{-\infty}^{\infty} f(x)dx = 1.$$

Discrete distributions do not admit a density, which is not too surprising, since a piecewise c.d.f. is non-differentiable.[2]

In the applications, it is not always obvious which type of the random variable to use for modelling. This can be illustrated by the following example.

Example 2.39 The tomorrow exchange rate for AUD/USD can be regarded as a random variable. In this case, we could accept the set of positive rational numbers as the set of possible values for the exchange rate. In this setting, the exchange rate will be represented by a discrete random variable. Alternatively, we could accept that the set of possible values is the set of all positive real numbers; the exchange rate will be represented by a continuous random variable. It can be noted that the continuous models are more common in financial modeling.

Remark 2.40 The set of the values of a random variable can be considered as the sample space for a probability model (compare Examples 1.3 and 2.39).

---

[2] There are continuous distributions that are not absolutely continuous, i.e. which do not admit a density. An example is a distribution supported on the so-called Cantor set known in classical Mathematical Analysis; it is a non-countable set with zero measure.

Functions of random variables

Let X be a random variable, and let $g : \mathbf{R} \to \mathbf{R}$ be a function. The transformation $Y = g(X)$ is also a random variable with some new distribution, and, possibly, with new properties. If the distribution of a random variable is known, then the distributions of all functions of this variable can also be derived.

Neither discrete nor continuous

There are random variables that are neither discrete nor continuous. In fact, it is true for all random variables such that their c.d.f. are neither piecewise constant nor continuous.

Example 2.41 Let X be a random variable distributed uniformly in $[0, 1]$. It is a continuous random variable. Let $Y = \max(1/3, X)$. Then Y is a random variable that is neither discrete nor continuous.

2.7 Examples of continuous random variables

Uniform distribution

Definition 2.42 *A random variable X with density*

$$f(x) = \begin{cases} 1, & x \in (0,1) \\ 0, & \text{otherwise} \end{cases}$$

is said to have the uniform distribution on the interval $[0, 1]$. We will write it as $X \sim U(0, 1)$.

For this distribution, all points in a finite interval are equally likely, and its c.d.f. is

$$F(x) = \begin{cases} 0, & x < 0, \\ x, & x \in (0,1) \\ 1, & x > 1. \end{cases}$$

In many programming languages, build-in random number generators produce the uniform distribution on $[0, 1]$ (in MATLAB, it is *rand* command).

Definition 2.43 *A random variable X with density*

$$f(x) = \begin{cases} (b-a)^{-1}, & x \in (a,b) \\ 0, & \textit{otherwise} \end{cases}$$

is said to have the uniform distribution on $[a,b]$ $(a < b)$. We will write it as $X \sim U(a,b)$.

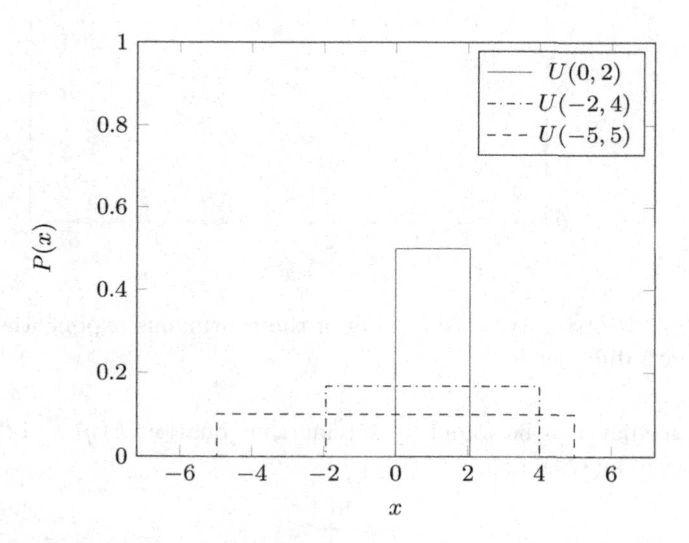

Figure 2.3: Probability density plots of the continuous uniform distribution given different intervals.

Theorem 2.44 *The median of the uniform distribution on $[0,1]$ is 0.5. The median of the uniform distribution on $[a,b]$ is $(b+a)/2$.*

Exponential distribution

Let $\lambda > 0$ be given. The exponential distribution has the density

$$f(x) = \begin{cases} \lambda e^{-\lambda x}, & x \geq 0 \\ 0, & \text{otherwise.} \end{cases}$$

Similarly to the Poisson distribution, the exponential density depends on a single parameter. The cumulative distribution function is

$$F(x) = \int_{-\infty}^{x} f(y)dy = \int_{0}^{x} f(y)dy = 1 - e^{-\lambda x}, \quad x > 0.$$

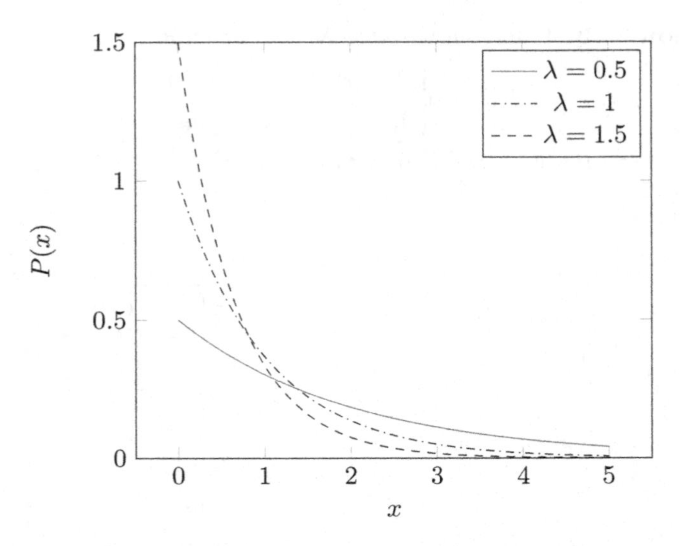

Figure 2.4: Probability density plots of the continuous exponential distribution with different λ.

The median can be found by solving the equation $F(\eta) = 1/2$. This gives $-\lambda\eta = \ln(1/2)$, or

$$\eta = \frac{\ln 2}{\lambda}.$$

The exponential distribution is often used to model lifetimes or waiting times, in which context it is conventional to replace x by t.

Varying λ corresponds to changing the units of measurement (for example, from seconds to minutes). Suppose that we consider modelling the lifetime T of an electronic component as an exponential random variable. We have:

$$\begin{aligned}\mathbf{P}(T > t + s | T > s) = \frac{\mathbf{P}(T > t + s, T > s)}{\mathbf{P}(T > s)} &= \frac{\mathbf{P}(T > t + s)}{\mathbf{P}(T > s)} \\ &= \frac{e^{-\lambda(t+s)}}{e^{-\lambda s}} = e^{-\lambda t}.\end{aligned}$$

We see that this probability does not depend on s. In this sense, the exponential distribution is *memoryless.* In particular, this means that it is not a very good model for human lifetimes, since the probability that a 20-year-old will live at least 10 more years is not the same as the probability that an 70-year-old will live will live at least 10 more years.

Gamma distribution

The Gamma distribution is a generalization of the exponential distribution. It has the density

$$f(x) = cx^{\alpha-1}e^{-\lambda x}, \quad t \geq 0.$$

If $x < 0$, then $f(x) = 0$. The coefficient $c > 0$ here is selected such that $\int_0^\infty f(x)dx = 1$.

If $\alpha = 1$ then it is an exponential density. The parameter α is called a shape parameter. Varying α changes the shape of the density, whereas varying λ corresponds to changing the units of measurement (say, from seconds to minutes) and does not affect the shape of the density.

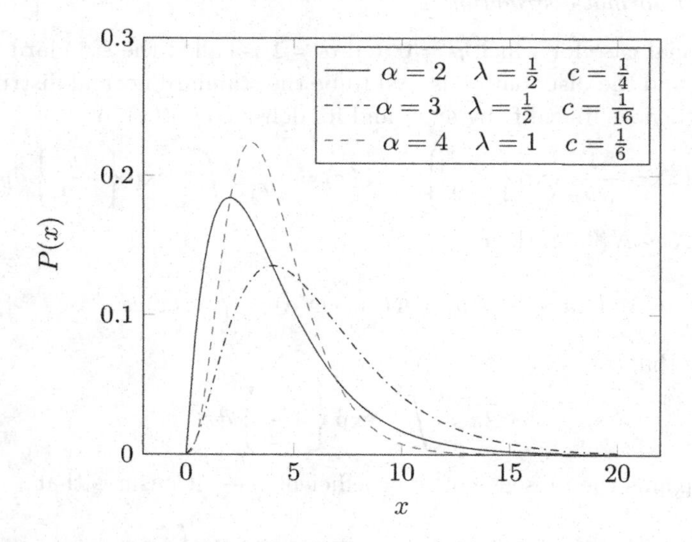

Figure 2.5: Probability density plots of the gamma distribution distributions with different shape(α) and scale(λ).

Normal or Gaussian distribution

Let $\sigma > 0$ and $\mu \in \mathbf{R}$ be given. A random variable X with the density

$$f(x) = \frac{1}{\sigma\sqrt{2\pi}} \exp\left\{-\frac{(x-\mu)^2}{2\sigma^2}\right\}$$

is said to be a Gaussian random variable with *mean* μ and *standard deviation* σ. It is said to have *the normal* or *Gaussian* distribution and is written as $X \sim N(\mu, \sigma^2)$.

That is an extremely important probability distribution which has applications in the statistics, probability theory, physics, engineering, finance, and economics. By *the Central Limit Theorem* which we will consider later, it is the limit distribution for mixing of a large number of independent random variables from a wide class of possible distributions.

The density is symmetric about μ, i.e., $f(x-\mu) = f(-(x-\mu))$, where it has a maximum, and that the rate at which it falls off is determined by σ.

The shapes of the normal densities are sometimes referred to as bell-shaped curves.

Standard normal distribution

The special case for which $\mu = 0$ and $\sigma = 1$ is called the standard normal density, and the distribution is said to be the standard normal distribution. We will denote its c.d.f. by $\Phi(x)$ and its density by $\phi(x)$, or

$$\phi(x) = \frac{1}{\sqrt{2\pi}} \exp\left\{-\frac{x^2}{2}\right\}, \quad \Phi(x) = \frac{1}{\sqrt{2\pi}} \int_{-\infty}^{x} \exp\left\{-\frac{y^2}{2}\right\} dy.$$

Let $X \sim N(0,1)$, then

$$\mathbf{P}(a \leq X \leq b) = \Phi(b) - \Phi(a) = \int_a^b \phi(x)dx.$$

Note that

$$\sqrt{2\pi} = \int_{-\infty}^{\infty} \exp\left\{-\frac{x^2}{2}\right\} dx.$$

This explains the presence of the coefficient $\frac{1}{\sqrt{2\pi}}$; it ensures that

$$\Phi(+\infty) - \Phi(-\infty) = \mathbf{P}(-\infty < X < +\infty) = \int_{-\infty}^{\infty} \phi(x)dx = 1.$$

The density is symmetric about $x = 0$, i.e., $\phi(x) = \phi(-x)$. In particular, it follows that

$$\Phi(+\infty) - \Phi(0) = \mathbf{P}(X > 0) = \Phi(0) - \Phi(-\infty) = \mathbf{P}(X < 0) = \frac{1}{2}.$$

In the scientific literature, it is also common to use $\Phi(x)$ for the so-called Laplace function

$$\frac{1}{\sqrt{2\pi}} \int_0^x \exp\left\{-\frac{y^2}{2}\right\} dy = \mathbf{P}(X \leq x) - 0.5$$

for $X \sim N(0,1)$, which is the c.d.f. for $N(0,1)$ minus 0.5.

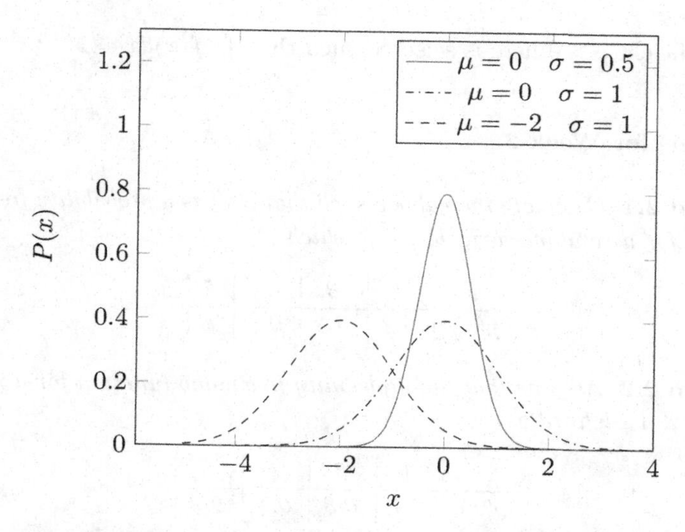

Figure 2.6: Probability density plots of normal distributions $N(\mu, \sigma^2)$.

Pareto distribution

We say that X has a Pareto distribution if it has the c.d.f.

$$F(x) = 1 - \left(\frac{\lambda}{\lambda + x}\right)^{\alpha}, \quad x \geq 0.$$

It can be rewritten as $X \sim Pa(\alpha, \lambda)$. We have that p.d.f. is

$$f(x) = \frac{\alpha\lambda^{\alpha}}{(\lambda + x)^{\alpha+1}}, \quad x \geq 0.$$

Generalized Pareto distribution

We say that X has a generalized Pareto distribution (or 3-parameter Pareto distribution) if it has the density

$$f(x) = c\frac{x^{k-1}}{(\lambda + x)^{\alpha+k}}, \quad x \geq 0.$$

The coefficient $c > 0$ here is selected such that $\int_0^{\infty} f(x)dx = 1$.

Beta distribution

Beta distribution B(a,b) has the density defined on the interval $[0, 1]$

$$f(u) = cu^{a-1}(1 - u)^{b-1}, \quad u \in [0, 1].$$

The coefficient $c > 0$ here is selected such that $\int_0^1 f(x)dx = 1$.

Problems for Week 2

Problem 2.1 *Calculate the value c such that $p(x)$ is a probability frequency function for a random variable X in which*

x	1	3	4	5
$p(x)$	0.2	0.3	0.4	c

Problem 2.2 *Assume that the probability frequency function for a random variable X is defined as*

x	1	3	4	5
$p(x)$	0.1	0.3	0.4	0.2

Let $F(x)$ be the c.d.f. of X. Find $F(-1)$, $F(1)$, $F(4.5)$.

Problem 2.3 *For X with the distribution described in the previous problem, find $\mathbf{P}(1 < X \leq 4.5)$.*

Problem 2.4 *Consider a casino roulette game. In this game, players make bets on certain colours. The probability that a red colour come up in one game is 18/38. Let X be the total number of the red colours in 5 games. Let $F(x)$ be the c.d.f. of X. Find $F(0.5)$.*

Problem 2.5 *Calculate the value k such that the function $f(x)$ is a probability density function. We assume that $f(x) = kx$ if $0 < x < 4$, and $f(x) = 0$ otherwise.*

Problem 2.6 *For X with the distribution described in the previous problem, find $\mathbf{P}(-2 < X < 1)$.*

Problem 2.7 *Let $X \sim U(0, 4)$. Find $\mathbf{P}(X^2 < 2)$.*

Problem 2.8 *Consider a casino roulette game. In this game, players make bets on certain numbers. The probability that the number 1 come up in one game is 1/38. Estimate the probability that this event happens at least once during 38 games. Use the Poisson distribution as an approximation.*

Problem 2.9 *Under the assumption of Problem 2.8, estimate the probability that the number* 1 *come up in one game is* $1/38$. *Estimate the probability that number* 1 *does not come up in entire sequence of* 76 *games.*

Problem 2.10 *Let* X *be uniformly distributed in* $[0,1]$, *and let* $Y = \min(X, 0.7)$. *(i) Find the c.d.f. for* Y. *Explain why* Y *is neither discrete nor continuous random variable. (ii) Find the c.d.f. for* $W = 2 - 2X$ *and* $Z = 3Y + 2$.

Problem 2.11 *The average number of emails arriving during any one hour at an email account is known to have a Poisson distribution, with the average number of emails known to be* 2. *What is the probability that during a two-hour period there will be no emails?*

Week 3. Joint Distributions

In this chapter, we describe the joint distributions of random variables, i.e., the distributions of random vectors.

3.1 The joint distribution of two random variables

Definition 3.1 *The joint distribution of two random variables X and Y (or bivariate distribution) is the set of probabilities*

$$\mathbf{P}(X \le x, \quad Y \le y)$$

for all $x \in \mathbf{R}$, $y \in \mathbf{R}$.

In other words, the joint distribution is defined by the cumulative distribution function

$$F(x, y) = \mathbf{P}(X \le x, Y \le y)$$

(or the bivariate c.d.f.).

A pair of random variables (X, Y) is said to be a random vector; the corresponding joint distribution can also be called the distribution of the vector. There is similar definition for $n \ge 2$ random variables.

Definition 3.2 *The joint distribution of n random variables (or multivariate distribution), $\{X_i\}_{i=1}^n$ is the set of probabilities*

$$\mathbf{P}(X_1 \le x_1, X_2 \le x_2, ..., X_n \le x_n),$$

for all $x_i \in \mathbf{R}$.

In other words, the joint distribution is uniquely defined by the multivariate cumulative distribution function

$$F(x_1, x_2, ..., x_n) = \mathbf{P}(X_1 \le x_1, ..., X_n \le x_n)$$

(the multivariate c.d.f.).

An ordered set of n random variables $(X_1, ..., X_n)$ can be considered as an n-dimensional random vector X with the components $(X_1, ..., X_n)$.

To make it more convenient for the methods of Linear Algebra, it is usually assumed that a random vector is a vector column, i.e., it is a mapping $X : \Omega \to \mathbf{R}^n$ rather then a mapping $X : \Omega \to \mathbf{R}^{n\times 1}$. Similarly, the random matrices can be defined as mappings $X : \Omega \to \mathbf{R}^{n\times m}$, or as matrices with random components.

3.2 Discrete random vectors

Suppose that X and Y are discrete random variables defined on the same sample space and that they take on values $x_1, x_2, ..., x_n, ...,$ and $y_1, y_2, ..., y_m, ...$ respectively, from some finite or countable sets. Their joint frequency function, or joint probability mass function, $p(x, y)$, is

$$p(x, y) = \mathbf{P}(X = x,\ Y = y).$$

It can also be considered as the distribution of the random vector (X, Y).

Suppose that we wish to find the frequency function of Y from the joint frequency function. We have

$$\begin{aligned}
&p(y_j) = \mathbf{P}(Y = y_j) \\
&= \mathbf{P}(Y = y_j, X = x_1) + \mathbf{P}(Y = y_j, X = x_2) + ... + \mathbf{P}(Y = y_j, X = x_n) \\
&= p(x_1, y_j) + p(x_2, y_j) + ... + p(x_n, y_j).
\end{aligned}$$

In other words, to find the frequency function of Y, we simply sum down the appropriate column of the table. For this reason, $p(y)$ is called the marginal frequency function of Y.

Similarly, summing across the rows gives

$$\begin{aligned}
&p(x_i) = \mathbf{P}(X = x_i) \\
&= \mathbf{P}(X = x_i, Y = y_1) + \mathbf{P}(X = x_i, Y = y_2) + ... + \mathbf{P}(X = x_i, Y = y_m) \\
&= p(x_i, y_1) + p(x_i, y_2) + ... + p(x_i, y_m),
\end{aligned}$$

which is the marginal frequency function of X.

Example 3.3 A fair coin is tossed three times; let X denote the number of heads on the first toss and Y denote the total number of heads. From the sample space

$$\Omega = \{hhh, hht, hth, htt, thh, tht, tth, ttt\},$$

we see that the joint frequency function of X and Y is as given in the following table:

$x\backslash y$	0	1	2	3
0	1/8	1/4	1/8	0
1	0	1/8	1/4	1/8

Hence

$$p_Y(0) = 1/8 + 0 = 1/8, \qquad p_X(1) = 1/8 + 1/4 + 1/8 = 1/2.$$

The case of several random variables

The case of several random variables is similar. If $X_1, ..., X_m$ are discrete random variables defined on the same sample space, their joint frequency function is

$$p(x_1, ...x_m) = \mathbf{P}(X_l = x_1, ..., X_m = x_m).$$

It can be interpreted also as the frequency function for the random vector $(X_1, ..., X_n)$. The marginal frequency function of X_1, for example, is

$$\begin{aligned} p(x_1) &= \mathbf{P}(X_1 = x_1) \\ &= \sum_{x_2, x_3, .., x_m} \mathbf{P}(X_1 = x_1, X_2 = x_2, X_3 = x_3, ..., X_m = x_m). \end{aligned}$$

3.3 Review of double integrals

We know

$$\int_a^b f(x)dx \sim \sum_i f(x_i)\Delta x_i$$

given that the interval (a, b) is the union of "small" disjoint intervals I_i such that $x_i \in I_i$ and

$$(a, b) = \cup_i I_i, \quad \text{mes}\,(I_i) = \Delta x_i.$$

We denote by mes the length of an interval. This approach can be generalized on two dimensional domains as follows.

Let $D = \{(x, y)\}$ be a domain in $\mathbf{R}^2$, and let $f(x, y)$ be a function. The integral $\int_D f(x, y)dxdy$ is

$$\int_D f(x, y)dxdy \sim \sum_i f(x_i, y_i)\Delta x_i \Delta y_i$$

given that D is approximated by the union of "small" disjoint rectangles D_i such that $(x_i, y_i) \in D_i$,

$$D \sim \cup_i D_i, \quad \text{mes}\,(D_i) = \Delta x_i \times \Delta y_i.$$

Here, we denote by mes the area of a domain. If $f(x, y) \equiv 1$ then

$$\int_D f(x,y)dxdy = \int_D dxdy = \text{mes}\,(D),$$

where mes (D) is the area of D.

Calculation: For the calculations, we replace the double integral by a repeating integral. Let

$$D = \{(x,y) : \ x \in (a,b), \ y \in (\alpha(x), \beta(x))\},$$

then

$$\int_D f(x,y)dxdy = \int_a^b dx \int_{\alpha(x)}^{\beta(x)} f(x,y)dy.$$

Similarly, if

$$D = \{(x,y) : \ y \in (c,d), \ x \in (\delta(y), \gamma(y))\},$$

then

$$\int_D f(x,y)dxdy = \int_c^d dy \int_{\delta(y)}^{\gamma(y)} f(x,y)dx.$$

Example 3.4 Let $f(x, y) \equiv 1$,

$$D = \{(x,y) : \ x \in (0,1), \ y \in (0,x)\}.$$

In this case,

$$\int_D f(x,y)dxdy = \int_D 1 \cdot dxdy = \int_0^1 dx \int_0^x dy = \int_0^1 dx \cdot x$$
$$= \frac{x^2}{2}\bigg|_0^1 = \frac{1}{2} = \text{mes}\,(D).$$

Note that, in the example above, the same D can be represented as

$$D = \{(x,y) : \ y \in (0,1), \ x \in (y,1)\}.$$

For $f(x, y) \equiv 1$, it gives again

$$\int_D f(x,y)dxdy = \int_D dxdy = \int_0^1 dy \int_y^1 dx = \int_0^1 dy \cdot (1-y)$$
$$= -\frac{(1-y)^2}{2}\bigg|_0^1 = \frac{1}{2} = \text{mes}\,(D).$$

3.4 Continuous random vectors

Suppose that X and Y are continuous random variables with a joint c.d.f. $F(x,y)$. Assume that there exists a function of two variables, $f(x,y)$, such that for any "reasonable" two-dimensional domain D,

$$P((X,Y)\in D)=\int_D f(x,y)dxdy.$$

We say that f is the joint density function of X and Y, or the joint probability density function (p.d.f.) of X and Y, or the p.d.f. of the vector (X,Y).

Clearly, any density must be such that

$$f(x,y)\geq 0,\quad \int_{-\infty}^{\infty}\int_{-\infty}^{\infty} f(x,y)dxdy=1.$$

It will be convenient to use so-called *indicator function* $\mathbb{I}_D$ defined for a set D by the following rule: $\mathbb{I}_D(x)=1$ if $x\in D$, and $\mathbb{I}_D(x)=0$ if $x\notin D$.

The cumulative distribution function (c.d.f.) $F(x,y)$ can be defined as

$$F(x,y)=\mathbf{P}((X\leq x,\ Y\leq y)=\int_{-\infty}^{x}\int_{-\infty}^{y} f(u,v)dvdu.$$

From the Fundamental Theorem of Calculus for the multivariable functions, it follows that

$$f(x,y)=\frac{\partial^2 F}{\partial x\partial y}(x,y).$$

Similarly to the case of discrete random variables, we obtain that the distribution of any random variable, X or Y (or so called marginal distributions), can be derived from the joint distribution:

$$\begin{aligned}F_X(x)=\mathbf{P}(X\leq x)&=\lim_{y\to+\infty}\mathbf{P}(X\leq x,Y\leq y)\\&=\lim_{y\to+\infty}F(x,y)=\int_{-\infty}^{x}du\int_{-\infty}^{\infty}f(u,v)dv,\end{aligned}$$

and

$$\begin{aligned}F_Y(y)=\mathbf{P}(Y\leq y)&=\lim_{x\to+\infty}\mathbf{P}(X\leq x,\ Y\leq y)\\&=\lim_{x\to+\infty}F(x,y)=\int_{-\infty}^{y}dv\int_{-\infty}^{\infty}f(u,v)du.\end{aligned}$$

In particular, it follows that the marginal densities can be found as

$$f_X(x)=\int_{-\infty}^{\infty}f(x,v)dv,\quad f_Y(y)=\int_{-\infty}^{\infty}f(u,y)du.$$

3.5 Uniform distribution for vectors

It will be convenient to use the so-called *indicator function* $\mathbb{I}_D$ defined for a set D by the following rule: $\mathbb{I}_D(x) = 1$ if $x \in D$ and $\mathbb{I}_D(x) = 0$ if $x \notin D$.

Definition 3.5 *Let $D \subset \mathbf{R}^2$ be a domain (region). Let (X, Y) be a random vector that has the probability density function*

$$f(x,y) = \text{const} \cdot \mathbb{I}_D(x,y).$$

We say that the vector (X, Y) is a random vector uniformly distributed in the domain D, or that (X, Y) is uniformly distributed in D.

The equation for the density above can be rewritten as

$$f(x,y) = \text{const}, \quad (x,y) \in D,$$
$$f(x,y) = 0, \quad (x,y) \notin D.$$

It can be shown that the constant here is uniquely defined from the equality $\int_{\mathbf{R}^2} f(x,y)dxdy = 1$. It can be rewritten as $\int_D f(x,y)dxdy = \text{const} \cdot \int_D dxdy = 1$. This gives

$$\text{const} = \left(\int_D dxdy\right)^{-1} = \frac{1}{\text{mes}\,(D)},$$

where mes denote the area of a domain.

Example 3.6 Let vector (X, Y) be uniformly distributed in the triangle

$$D = \{(x,y) : \ x \in (0,1), \ y \in (0,x)\}.$$

Let us find the marginal densities.

Solution. We have that $\text{mes}\,(D) = 1/2$. Hence the joint density is

$$f(x,y) = (1/2)^{-1}\mathbb{I}_D(x,y) = 2\mathbb{I}_D(x,y).$$

Further, we have that

$$f_X(x) = \int_{-\infty}^{\infty} f(x,y)dy = \int_0^x 2dy = 2x, \quad x \in [0,1];$$

and

$$f_Y(y) = \int_{-\infty}^{\infty} f(x,y)dx = \int_y^1 2dx = 2(1-y), \quad y \in [0,1].$$

It can be seen that the marginal densities are not uniform in this example. Let us verify that $\int_{-\infty}^{\infty} f_X(x)dx = 1$. We have

$$\int_{-\infty}^{\infty} f_X(x)dx = \int_0^1 2xdx = 2\frac{x^2}{2}\bigg|_0^1 = 1.$$

3.6 Several continuous random variables

We can make the obvious generalizations. The joint density function is a function of several variables, and the marginal density functions are found by integration. There are marginal density functions of various dimensions. Suppose that X, Y, Z are jointly continuous random variables with the density function $f(x, y, z)$. The one-dimensional marginal distribution of X is

$$f_X(x) = \int_{-\infty}^{\infty} \int_{-\infty}^{\infty} f(x, y, z) dy dz,$$

and the two-dimensional marginal distribution of (X, Y) is

$$f_{XY}(x, y) = \int_{-\infty}^{\infty} f(x, y, z) dz.$$

Example 3.7 One can define random vectors (X, Y, Z) uniformly distributed in a domain $D \subset \mathbf{R}^3$. In this case, the density is

$$f(x, y, z) = \frac{1}{\text{mes}\,(D)} \mathbb{I}_D(x, y, z),$$

where mes denote the volume of a domain in $\mathbf{R}^3$. This definition can be extended on domains in $\mathbf{R}^n$ for any n.

3.7 Bivariate normal density

The definition of a normal random variable can be extended on the multi-dimensional case.

Let us consider the two-dimensional case. We say that a pair of random variables X and Y has a bivariate normal density if their joint density is given as

$$f(x, y) = \frac{1}{2\pi\sigma_X\sigma_Y\sqrt{1-\rho^2}}$$
$$\times \exp\left(-\frac{1}{2(1-\rho^2)}\left[\frac{(x-\mu_X)^2}{\sigma_X^2} + \frac{(y-\mu_Y)^2}{\sigma_Y^2} - \frac{2\rho(x-\mu_X)(y-\mu_Y)}{\sigma_X\sigma_Y}\right]\right).$$

The density depends on 5 parameters:

$$\mu_X, \mu_Y \in \mathbf{R}, \quad \sigma_X > 0, \quad \sigma_Y > 0, \quad \rho \in (-1, 1).$$

Note that the curves of $\mathbf{R}^2$ where $f(x, y) = \text{const}$ are given by circles if $\rho = 0$ or by elliptic curves if $\rho \neq 0$ (if $\sigma_X = \sigma_Y$). The center of any ellipse or circle is at the point (μ_X, μ_Y).

Theorem 3.8 *The marginal densities for the bivariate normal density are* $N(\mu_X, \sigma_X^2)$ *and* $N(\mu_Y, \sigma_Y^2)$.

3.8 The distribution of independent random variables

Remind that random variables X and Y are said to be independent if, for all $a_1, b_1, a_2, b_2 \in \mathbf{R}$,

$$\mathbf{P}(X \in (a_1, b_1] \text{ and } Y \in (a_1, b_2]) = \mathbf{P}(X \in (a_1, b_1])\mathbf{P}(Y \in (a_2, b_2]).$$

It can be shown that this definition is equivalent to the following definition.

Definition 3.9 *Two random variables* X *and* Y *are said to be independent if, for all* $x, y \in \mathbf{R}$,

$$F(x, y) = F_X(x)F_Y(y).$$

Here $F(x, y)$ *is the joint c.d.f. for* (X, Y), *and* F_X *and* F_Y *are the marginal c.d.f.*

It follows that, for independent discrete random variables, the joint frequency can be represented as the product of the marginal frequencies

$$p(x, y) = p_X(x)p_Y(y).$$

Similarly, for independent continuous random variables, the joint density can be represented as the product of marginal densities

$$f(x, y) = f_X(x)f_Y(y).$$

These definitions can be extended on the case of joint distribution of $n \geq 2$ random variables.

Definition 3.10 *Random variables* $X_1, ..., X_n$ *are said to be independent if, for all* $x_i \in \mathbf{R}$,

$$F(x_1, ..., x_n) = F_{X_1}(x_1) \cdots F_{X_n}(x_n).$$

Here $F(x_1, ..., x_n)$ *is the joint c.d.f., and* F_{X_i} *are the marginal c.d.f.'s.*

Example 3.11 If X and Y follow a bivariate normal distribution with $\rho = 0$, their joint density factors into the product of two normal densities. Therefore X and Y are independent.

Example 3.12 Let a random variable X be uniformly distributed on an interval $[a_1, b_1]$ and let Y be uniformly distributed on an interval $[a_2, b_2]$. Assume that X and Y are independent. Prove that the random vector (X, Y) is uniformly distributed on $[a_1, b_1] \times [a_2, b_2]$.

Solution. The random variable X has density

$$f_X(x) = \text{const} \cdot \mathbb{I}_{[a_1,b_1]}(x).$$

The random variable Y has density

$$f_Y(y) = \text{const} \cdot \mathbb{I}_{[a_2,b_2]}(y).$$

It follows from independence assumption that the random vector (X, Y) has a density $f(x, y) = f_X(x)f_Y(y)$. Hence

$$f(x, y) = \text{const} \cdot \mathbb{I}_{[a_1,b_1]\times[a_2,b_2]}(x, y).$$

By the definition, this vector is uniformly distributed on the corresponding rectangle.

3.9 Mixed case

Let $X \sim U(0, 1)$ (uniform on $[0, 1]$) and Y has Bernoulli distribution with parameter $p = 1/3$ (i.e. $\mathbf{P}(Y = 1) = 1/3$, $\mathbf{P}(Y = 0) = 2/3$). Then the joint distribution is neither discrete nor continuous. If X and Y are independent then the joint c.d.f. is

$$F(x, y) = F_X(x)F_Y(y).$$

3.10 Conditional frequencies

Consider discrete random variables X and Y with the joint frequency $p(x, y)$ and with the marginal frequencies $p_X(x)$ and $p_Y(y)$. By the definition of conditional probability, we have

$$\mathbf{P}(X = x|Y = y) = \frac{\mathbf{P}(X = x, Y = y)}{\mathbf{P}(Y = y)}.$$

Therefore, the conditional frequency function can be defined as

$$p_{X|Y}(x|y) = \frac{p(x, y)}{p_Y(y)}.$$

This probability is defined to be zero if $p_Y(y) = 0$.

Note that this function is a genuine frequency function since it is non-negative and sums to 1.

It can also be seen that if X and Y are independent then $p_{X|Y}(x|y) = p_X(x)$. In this case, the knowledge of the value of Y does not change our estimates for probabilities related to values of X.

It follows from these definitions that

$$p(x,y) = p_{X|Y}(x|y)p_Y(y),$$

and

$$p_X(x) = \sum_y p_{X|Y}(x|y)p_Y(y).$$

Example 3.13 Consider again the example when a fair coin is tossed three times; X is the number of heads on the first toss and Y is the total number of heads. The joint frequency function $p(x,y)$ of X and Y is given in the following table:

$x\backslash y$	0	1	2	3
0	1/8	1/4	1/8	0
1	0	1/8	1/4	1/8

We have that $p_X(0) = p_X(1) = 1/2$ and $p_{Y|X}(y|x) = p(x,y)/p(x)$. It gives the table for $p_{Y|X}(y|0) = p(0,y)/p(0)$

$x\backslash y$	0	1	2	3
0	1/4	1/2	1/4	0

Similarly, the table for $p_{Y|X}(y|1) = p(1,y)/p(0)$ is

$x\backslash y$	0	1	2	3
1	0	1/4	1/2	1/4

3.11 Conditional densities

Let X and Y be two continuous random variables with the joint density f and with the marginal densities f_X and f_Y. By the definition of conditional probability, we have, for small dx and dy, that

$$\mathbf{P}(X \in (x, x+dx) | Y \in (y, y+dy)) = \frac{\mathbf{P}(X \in (x, x+dx), Y \in (y, y+dy))}{\mathbf{P}(Y \in (y, y+dy))}$$

$$= \frac{\int_x^{x+dx} \int_y^{y+dy} f(u,v)dvdu}{\int_y^{y+dy} f_Y(v)dv} \approx \frac{f(x,y)dxdy}{f_Y(y)dy}.$$

This value is proportional to dx. Therefore, the conditional density function can be defined as

$$f_{X|Y}(x|y) = \frac{f(x,y)}{f_Y(y)}.$$

It can also be rewritten as

$$f_{X|Y}(x|y) = f_X(x|Y = y),$$

meaning that $f_X(x|Y = y)$ is the p.d.f. for X under the conditional probability given that $Y = y$. This means, for any interval $A \subset \mathbf{R}$, we have

$$\mathbf{P}(X \in A|Y = y) = \int_A f_X(x|Y = y)dx.$$

Note that $f_{X|Y}(y|x) = f_X(x)$ if X and Y are independent. In this case, the knowledge of the value of Y does not change our estimates for probabilities related to values of X.

3.12 Bayes' formula for the densities

It follows from the definitions that

$$f(x,y) = f_{X|Y}(x|y)f_Y(y).$$

Hence

$$f_X(x) = \int_{-\infty}^{+\infty} f(x,y)dy = \int_{-\infty}^{+\infty} f_{X|Y}(x|y)f_Y(y)dy.$$

The following theorems give us analogs of Bayes' formula for continuous distributions.

Theorem 3.14 *Let $A \subset \mathbf{R}$ be an interval, let (X,Y) be a random vector with the joint density $f(x,y)$, and Y has the marginal density $f_Y(y)$. Then*

$$\mathbf{P}(X \in A) = \int_{-\infty}^{\infty} \mathbf{P}(X \in A|Y = y)f_Y(y)dy.$$

Proof. We have

$$\mathbf{P}(X \in A) = \int_A f_X(x)dx = \int_A dx \int_{-\infty}^{\infty} f_{X|Y}(x|y)f_Y(y)dy$$

$$= \int_{-\infty}^{\infty} f_Y(y)dy \int_A f_{X|Y}(x|y)dx = \int_{-\infty}^{\infty} f_Y(y)\mathbf{P}(X \in A|Y = y).$$

Hence the proof follows.

Theorem 3.15 *Let $D \subset \mathbf{R}^2$ be a domain, (X, Y) be a random vector with joint density $f(x,t)$ and Y has marginal density $f_Y(y)$. Then*

$$\mathbf{P}((X,Y) \in D) = \int_{-\infty}^{\infty} \mathbf{P}((X,Y) \in D | Y = y) f_Y(y) dy.$$

Proof. Let $D(y) = \{x : (x,y) \in D\}$. We have

$$\mathbf{P}((X,Y) \in D) = \int_D f(x,y) dx dy = \int_D f_{X|Y}(x|y) f_Y(y) dx dy$$
$$= \int_{-\infty}^{\infty} f_Y(y) dy \int_{D(y)} f_{X|Y}(x|y) dx.$$

Further, we have

$$\mathbf{P}((X,Y) \in D | Y = y) = \mathbf{P}(X \in D(y) | Y = y) = \int_{D(y)} f_{X|Y}(x|y) dx.$$

Thus the proof follows.

Example 3.16 Consider again the vector (X, Y) uniformly distributed in the triangle

$$D = \{(x,y) : x \in (0,1), \ y \in (0,x)\}.$$

Let us find the conditional densities.

Solution. We have that the joint density is

$$f(x,y) = 2\mathbb{I}_D(x,y),$$

and that the marginal densities are

$$f_X(x) = 2x, \quad x \in [0,1], \qquad f_Y(y) = 2(1-y), \quad y \in [0,1].$$

Hence

$$f_{Y|X}(y|x) = \frac{f(x,y)}{f_X(x)} = \mathbb{I}_D(x,y)\frac{2}{2x} = \mathbb{I}_D(x,y)\frac{1}{x},$$

and

$$f_{X|Y}(x|y) = \frac{f(x,y)}{f_Y(y)} = \mathbb{I}_D(x,y)\frac{2}{2(1-y)} = \mathbb{I}_D(x,y)\frac{1}{1-y}.$$

It follows that $f_{Y|X}(y|x) \to +\infty$ as $x \to 0$ and $f_{X|Y}(x|y) \to +\infty$ as $y \to 1$ (there is an explanation for this). It can also be seen that the conditional distribution is a uniform one.

Problems for Week 3

Problem 3.1 *Suppose that the joint density for X and Y is defined as $f(x,y) = ky$, $(x,y) \in D$, $f(x,y) = 0$, $(x,y) \notin D$. Here $k \in \mathbf{R}$, $D = \{(x,y):\ x \in [0,1],\ y \in [0,1]\}$. Find k and find $\mathbf{P}(Y < 1/2)$.*

Problem 3.2 *Consider a joint density for a random vector (X,Y) defined as $f(x,y) = ky$, $(x,y) \in D$, $D = \{(x,y):\ x \in [0,1],\quad y \in [-1,0]\}$. Find k and find $\mathbf{P}(Y < -1/2)$.*

Problem 3.3 *Let X and Y be independent random variables. Let X be uniformly distributed on the interval $(-1,1)$, and let*

$$Y = \begin{cases} 0.3, & \text{with probability } 1/5 \\ 0.7, & \text{with probability } 4/5. \end{cases}$$

Find $\mathbf{P}(Y + |X| < 1)$. (Hint: you may use the Bayes' formula.)

Problem 3.4 *Let $D = \{(x,y) : 0 \le x \le 1,\ 0 \le y \le 1 - x\}$. Let two random variables X and Y be such that the random vector (X,Y) is uniformly distributed on D.*

(a) Find the joint density $f(x,y)$, the marginal densities $f_X(x)$ and $f_Y(y)$, and the conditional density $f_{X|Y}(x|y)$.
(b) Find $\mathbf{P}(X + Y < 0.5)$.
(c) Find $\mathbf{P}(Y > 0.5 \,|\, X = 0.5)$.
(d) Find $\mathbf{P}(0 < Y < 1 - X) = 1$.

Problem 3.5 *Let X and Y be independent random variables. Assume that X has the density*

$$f(x) = \begin{cases} ce^{-2x}, & x \ge 1 \\ 0, & \text{otherwise}, \end{cases}$$

where $c > 0$ is a constant. Assume that Y is uniformly distributed in the interval $(0,2)$.

(a) Find c.
(b) Find the median of the distribution for X.
(c) Find the joint density $f(x,y)$ of (X,Y).
(d) Using (c), find $\mathbf{P}(X < Y)$. (Hint: sketch the corresponding domains first.)

Week 4. Transformations of the Distributions

In this chapter, we describe methods of estimating the distributions of functions of random variables and vectors.

4.1 The density for the invertible functions

Let X have density $f(x)$, and let

$$U = g(X)$$

be a new random variable. In this section, we consider the case where the function can be inverted, i.e., that there exists functions $h : \mathbf{R} \to \mathbf{R}$ such that

$$X = h(U).$$

Example 4.1 There are functions that cannot be inverted; for instance, the function $U = |X|$ is not invertible since X cannot be represented as a function of U.

Example 4.2 The following functions are invertible: $U = a + bX$, $b \neq 0$, $U = X^3$, or $U = e^{2X}$, etc.

Jacobian for a function

Let

$$J(u) = \frac{dh}{du}(u).$$

This function is called the Jacobian for the function h. We assume that the derivative here exists and that $J(u) \neq 0 \quad \forall u$. ($\forall$ is the standard mathematical notation for "for all".)

Theorem 4.3 *Assume a random variable X has the density $f_X(x)$. The density for U is then*

$$f_U(u) = f_X(h(u))|J(u)|.$$

Proof. Assume that $\dfrac{dh(u)}{du} > 0$. We have that

$$\mathbf{P}(U \le u) = \mathbf{P}(g(X) \le u) = \mathbf{P}(X \le h(u)) = \int_{-\infty}^{h(u)} f_X(x)dx.$$

By differentiating, we obtain

$$f_U(u) = \frac{\mathbf{P}(U \le u)}{du} = f_X(h(u))\frac{dh}{du}(u) = f_X(h(u))|J(u)|.$$

Now, let us assume that $\dfrac{dh(u)}{du} < 0$. Similarly,

$$\mathbf{P}(U \le u) = \mathbf{P}(g(X) \le u) = \mathbf{P}(X \ge h(u)) = \int_{h(u)}^{+\infty} f_X(x)dx.$$

By differentiating again, we get

$$f_U(u) = \frac{\mathbf{P}(U \le u)}{du} = -\frac{dh}{du}(g(x))f_X(h(u)) = f_X(h(u))|J(u)|.$$

Corollary 4.4 *If X has the probability density function $f_X(x)$, then $U = a + bX$ has the density function*

$$f_U(u) = |b|^{-1} f_X\left(\frac{u-a}{b}\right).$$

Proof. We apply Theorem 4.3 with $U = g(X)$, $g(x) = a + bx$, $h(u) = (u-a)/b$, and $J(u) = b^{-1}$.

Connection with the Jacobian of g

Let us observe that $h(g(x)) = x$ and

$$1 = \frac{d}{dx}h(g(x)) = \frac{dh}{du}(g(x))\frac{dg}{dx}(x).$$

Hence

$$\frac{dh}{du}(g(x)) = \left(\frac{dg}{dx}(x)\right)^{-1}.$$

By the definitions,

$$\frac{dh}{du}(u) = \frac{dh}{du}(g(h(u))\left(\frac{dg}{dx}(h(u))\right)^{-1} = J_g(h(u))^{-1},$$

where J_g is the corresponding Jacobian of the function g. Sometimes, it is convenient to use J_g instead of J. It may happen, for instance, when the derivative of J_g is easier to derive.

Example 4.5 Let $X \sim U(0,1)$. Therefore $f_X(x) = \mathbb{I}_{[0,1]}(x)$. Assume $Y = X + 3$. Thus, the density for Y is

$$f_Y(x) = \mathbb{I}_{[0,1]}(x-3) = \mathbb{I}_{[3,4]}(x).$$

Example 4.6 Let $X \sim U(0,1)$. Therefore $f_X(x) = \mathbb{I}_{[0,1]}(x)$. Assume $Y = 2X$. Thus, the density for Y is

$$f_Y(x) = \frac{1}{2}\mathbb{I}_{[0,1]}(x/2) = \frac{1}{2}\mathbb{I}_{[0,2]}(x).$$

Example 4.7 Let $X \sim U(0,1)$. Therefore $f_X(x) = \mathbb{I}_{[0,1]}(x)$. Assume $Y = 2X + 3$. Thus, the density for Y is

$$f_Y(x) = \frac{1}{2}\mathbb{I}_{[0,1]}((x-3)/2) = \frac{1}{2}\mathbb{I}_{[3,5]}(x).$$

Example 4.8 Let $X \sim U(0,1)$. Therefore $f_X(x) = \mathbb{I}_{[0,1]}(x)$. Assume $Y = -2X + 3$. Thus, the density for Y is

$$f_Y(x) = \frac{1}{2}\mathbb{I}_{[0,1]}((x-3)/(-2)) = \frac{1}{2}\mathbb{I}_{[0,1]}((3-x)/2) = \frac{1}{2}\mathbb{I}_{[-1,1]}(x).$$

Example 4.9 Let X have exponential distribution $X \sim Exp(\lambda)$. Assume $U = X/2$. To find the density for U, let $g(x) = x/2$, $h(u) = 2u$. Therefore, the Jacobian is

$$J(u) = \frac{dh}{du}(u) = 2.$$

Hence, the density for U is

$$f_U(u) = f_X(h(u))|J(u)| = f_X(h(u))2 = 2\lambda \exp[-\lambda(2u)]\mathbb{I}_{\{u>0\}}(u),$$

which is also an exponential distribution.

Direct calculation of the densities

In many cases, we have to calculate the c.d.f.'s for transformed random variables directly. Similarly, we can find the densities for the transformed variables via recalculation of the c.d.f.

Example 4.10 Assume X has the density $f_X(x)$ and $Y = |X|$. Let us find the c.d.f. $F_Y(y)$ and the density $f_Y(y)$ for Y.

Solution. For $y \le 0$, we have $F_Y(y) = 0$. For $y > 0$, we have

$$F_Y(y) = \mathbf{P}(Y \le y) = \mathbf{P}(|X| \le y) = \mathbf{P}(-y \le X \le y) = \mathbf{P}(-y < X \le y)$$
$$= F_X(y) - F_X(-y).$$

By differentiating, we obtain the corresponding densities:
For $y \leq 0$, $f_Y(y) = 0$. For $y > 0$,

$$f_Y(y) = \frac{d}{dy} F_Y(y) = \frac{d}{dy}(F_X(y) - F_X(-y))$$
$$= f_X(y) + f_X(-y).$$

4.2 Sums of random variables

Sums of discrete random variables

Let X and Y be discrete random variables with joint frequency function $p(x, y)$ and $Z = X + Y$. Find the frequency function $p_Z(z)$ of Z.
Assume that X and Y only take integer values. We have

$$Z = z \quad \text{iff (if and only if)} \quad X = x,\ Y = z - x.$$

Hence

$$p_Z(z) = \mathbf{P}(Z = z) = \sum_x \mathbf{P}(X = x,\ Y = z - x) = \sum_x p(x, z - x).$$

If X and Y are independent, then

$$p_Z(z) = \mathbf{P}(Z = z) = \sum_x p_X(x) p_Y(z - x).$$

In this case, this sum is called the *convolution* of the sequences $p_X(x)$ and $p_Y(y)$.

Example 4.11 A fair coin is tossed three times. Let X denote the number of heads on the first toss and Y denote the total number of heads. We found already that the joint frequency function $p(x, y)$ is given as

$x \backslash y$	0	1	2	3
0	1/8	1/4	1/8	0
1	0	1/8	1/4	1/8

Let $Z = X + Y$. Then

$$p_Z(1) = \sum_{x=0,1} p(x, 1-x) = p(0, 1-0) + p(1, 1-1)$$
$$= p(0,1) + p(1,0) = 1/4,$$
$$p_Z(2) = \sum_{x=0,1} p(x, 2-x) = p(0, 2-0) + p(1, 2-1)$$
$$= p(0,2) + p(1,1) = 1/8 + 1/8 = 1/4,$$
$$p_Z(3) = \sum_{x=0,1} p(x, 3-x) = p(0, 3-0) + p(1, 3-1)$$
$$= p(0,3) + p(1,2) = 1/4,$$
$$p_Z(4) = \sum_{x=0,1} p(x, 4-x) = p(0, 4-0) + p(1, 4-1)$$
$$= p(0,4) + p(1,3) = 0 + 1/8.$$

Sums of continuous random variables

Let X and Y be continuous random variables and $Z = X + Y$. We have that

$Z \leq z$ if and only if the vector (X, Y) is in the region $\{x + y \leq z\}$.

Hence, the c.d.f. of Z is

$$F_Z(z) = \mathbf{P}(Z \leq z)) = \int_{(x,y):\ x+y\leq z} f(x,y)dxdy$$
$$= \int_{-\infty}^{\infty} dx \int_{-\infty}^{z-x} f(x,y)dy.$$

Differentiating, we obtain

$$f_z(z) = \frac{dF_Z}{dz}(z) = \frac{d}{dz}\int_{-\infty}^{\infty} dx \int_{-\infty}^{z-x} f(x,y)dy.$$

We can calculate this using the rule for *differentiation under the integral sign:*

$$\frac{d}{dz}\int_a^b g(x,z)dx = \int_a^b \frac{dg}{dz}(x,z)dx.$$

It gives

$$f_z(z) = \int_{-\infty}^{\infty} dx \frac{d}{dz}\int_{-\infty}^{z-x} f(x,y).$$

By the Fundamental Theorem of Calculus and by the Chain Rule, we have

$$\frac{d}{dz}\int_{-\infty}^{z-x} f(x,y)dy = f(x, z-x).$$

Hence

$$f_Z(z) = \int_{-\infty}^{\infty} f(x, z-x)dx.$$

It is the analogue of the result for the discrete case.

If X and Y are independent, then

$$f_z(z) = \int_{-\infty}^{\infty} f_X(x)f_Y(z-x)dx.$$

This integral is also called the *convolution* of the functions f_X and f_Y.

Example 4.12 Let T_1 and T_2 be independent exponential random variables with the same density $f_T(t) = \lambda e^{-\lambda t}$, $t \geq 0$. Find the density for

$$S = T_1 + T_2.$$

Solution. The density for S is

$$\begin{aligned} f_S(s) &= \int_{-\infty}^{\infty} f_T(t)f_T(s-t)dt = \int_0^s f_T(t)f_T(s-t)dt \\ &= \lambda^2 \int_0^s e^{-\lambda t}e^{-\lambda(s-t)}dt = \lambda^2 \int_0^s e^{-\lambda s}dt \\ &= \lambda^2 e^{-\lambda s} \textstyle\int_0^s dt = \lambda^2 s e^{-\lambda s}. \end{aligned}$$

This is a Gamma distribution.

4.3 Ratio distribution/quotients

Ratio of discrete random variables

Let X and Y be discrete random variables with integer values with joint frequency function $p(x,y)$, $\mathbf{P}(X=0) = 0$ and $Z = Y/X$. Let us find the frequency function $p_Z(z)$ of Z.

We have

$$Z = z \quad \text{if and only if} \quad X = x, \quad Y = zx.$$

Hence

$$\mathbf{P}(Z=z) = \sum_x p(x, zx).$$

If X and Y are independent, then

$$\mathbf{P}(Z=z) = \sum_x p_X(x)p_Y(zx).$$

Example 4.13 A fair coin is tossed three times. Let X_0 denote the number of heads on the first toss, $X = X_0 + 1$ and Y be the total number of heads. We found above that the joint frequency function $p_{X_0 Y}(x, y)$ is given by:

$x \backslash y$	0	1	2	3
0	1/8	1/4	1/8	0
1	0	1/8	1/4	1/8

It follows that the frequency function $p(x, y)$ for (X, Y) is:

$x \backslash y$	0	1	2	3
1	1/8	1/4	1/8	0
2	0	1/8	1/4	1/8

Let us find the probability frequency function for $Z = Y/X$. We have that

$$p_Z(0) = \sum_{x=1,2} p(x, 0 \cdot x) = p(1,0) + p(2,0) = 1/8 + 0 = 1/8,$$

$$\begin{aligned} p_Z(1) &= \sum_{x=1,2} p(x, 1 \cdot x) = p(1, 1 \cdot 1) + p(2, 1 \cdot 2) \\ &= p(1,1) + p(2,2) = 1/4 + 1/4 = 1/2, \end{aligned}$$

$$\begin{aligned} p_Z(0.5) &= \sum_{x=1,2} p(x, 0.5 \cdot x) = p(1, 0.5 \cdot 1) + p(2, 0.5 \cdot 2) \\ &= p(1, 0.5) + p(2,1) = 0 + 1/8 = 1/8, \end{aligned}$$

$$\begin{aligned} p_Z(1.5) &= \sum_{x=1,2} p(x, 1.5 \cdot x) = p(1, 1.5 \cdot 1) + p(2, 1.5 \cdot 2) \\ &= p(1, 1.5) + p(2,3) = 0 + 1/8 = 1/8, \end{aligned}$$

and

$$\begin{aligned} p_Z(2) &= \sum_{x=1,2} p(x, 2 \cdot x) = p(1, 2 \cdot 1) + p(2, 2 \cdot 2) \\ &= p(1,2) + p(1,4) = 1/8 + 0 = 1/8, \end{aligned}$$

$$\begin{aligned} p_Z(3) &= \sum_{x=1,2} p(x, 3 \cdot x) = p(1, 3 \cdot 1) + p(2, 3 \cdot 2) \\ &= p(1,3) + p(1,6) = 0. \end{aligned}$$

Ratio of continuous random variables

Let X and Y be continuous random variables. Let us find the c.d.f. of $Z = Y/X$. We have that

$$Z \leq z \quad \text{iff} \quad (X, Y) \text{ is in the region } \{y/x \leq z\}.$$

If $x > 0$, this is the set $y < xz$. If $x < 0$, it is the set $y > xz$. Thus

$$\begin{aligned} F_Z(z) = \mathbf{P}(Z \leq z) &= \int_{(x,y):\ y/x \leq z} f(x,y)dxdy \\ &= \int_{-\infty}^{0} dx \int_{zx}^{+\infty} f(x,y)dy + \int_{0}^{\infty} dx \int_{-\infty}^{zx} f(x,y)dy. \end{aligned}$$

The density of Z is

$$f_z(z) = \frac{dF_Z}{dz}(z) = \frac{d}{dz} \int_{-\infty}^{0} dx \int_{zx}^{+\infty} f(x,y)dy + \frac{d}{dz} \int_{0}^{\infty} dx \int_{-\infty}^{zx} f(x,y)dy.$$

We can calculate it using again the *differentiation under the integral sign* rule; it gives

$$f_z(z) = \int_{-\infty}^{0} dx \frac{d}{dz} \int_{zx}^{+\infty} f(x,y)dy + \int_{0}^{\infty} dx \frac{d}{dz} \int_{-\infty}^{zx} f(x,y)dy.$$

By the Fundamental Theorem of Calculus and by Chain Rule for differentiation,

$$\frac{d}{dz} \int_{zx}^{+\infty} f(x,y)dy = -xf(x,zx), \qquad \frac{d}{dz} \int_{-\infty}^{zx} f(x,y)dy = xf(x,zx).$$

Hence

$$f_Z(z) = -\int_{-\infty}^{0} xf(x,zy)dx + \int_{0}^{\infty} xf(x,zx)dx.$$

We have that $-x = |x|$ for $x < 0$ and $|x| = x$ for $x > 0$. Therefore

$$f_Z(z) = \int_{-\infty}^{\infty} dx|x|f(x,zx)dx.$$

If X and Y are independent,

$$f_z(z) = \int_{-\infty}^{\infty} |x|f_X(x)f_Y(zx)dx.$$

Cauchy density

Suppose that X and Y are independent standard normal random variables $N(0,1)$. Let $Z = Y/X$. By the definitions of $N(0,1)$, we have densities for X and Y as

$$f_X(x) = \frac{1}{\sqrt{2\pi}} e^{-\frac{x^2}{2}}, \quad f_Y(y) = \frac{1}{\sqrt{2\pi}} e^{-\frac{y^2}{2}}.$$

Then

$$\begin{aligned} f_z(z) &= \int_{-\infty}^{\infty} |x| f_X(x) f_Y(zx) dx = \int_{-\infty}^{\infty} |x| f_X(x) f_Y(zx) dx \\ &= \int_{-\infty}^{\infty} |x| \frac{1}{\sqrt{2\pi}} \exp\left\{-\frac{x^2}{2}\right\} \frac{1}{\sqrt{2\pi}} \exp\left\{-\frac{z^2x^2}{2}\right\} dx \\ &= \frac{1}{2\pi} \int_{-\infty}^{\infty} |x| \exp\left\{-\frac{x^2 + z^2x^2}{2}\right\} dx. \end{aligned}$$

From the symmetry of the integrand about 0, we have

$$f_z(z) = \frac{1}{\pi} \int_0^{\infty} x \exp\left\{-\frac{x^2 + z^2x^2}{2}\right\} dx.$$

To simplify this, let $u = x^2$. Hence, $du = 2xdx$ or $xdx = du/2$. We obtain

$$f_z(z) = \frac{1}{\pi} \int_0^{\infty} \exp\left\{-\frac{u + z^2u}{2}\right\} \frac{1}{2} du = \frac{1}{\pi(1+z^2)}.$$

This density is called the *Cauchy density.*

It can be noted that the Cauchy density is bell-shaped, but the tails of the Cauchy tend to zero very slowly compared to the tails of the normal.

4.4 Transformation of the joint density

Let X and Y have the joint density $f(x,y)$, $g_1 : \mathbf{R}^2 \to \mathbf{R}$ and $g_2 : \mathbf{R}^2 \to \mathbf{R}$ be two functions. Assume that

$$U = g_1(X,Y), \quad V = g_2(X,Y)$$

are two new random variables.

Example 4.14 There are many examples of transformations

$$U = X + Y, \quad V = X - 2Y,$$

or

$$U = e^{2X}, \quad V = e^Y,$$

etc.

We assume that the transformation can be inverted, i.e., that there exists functions $h_1 : \mathbf{R}^2 \to \mathbf{R}$ and $h_2 : \mathbf{R}^2 \to \mathbf{R}$ such that

$$X = h_1(U, V), \quad Y = h_2(U, V).$$

Example 4.15 There are transformations that cannot be inverted; for instance, the transformation $U = X$, $V = |Y|$ is not invertible since Y cannot be represented as a function of V.

Definition 4.16 *Consider a matrix:*

$$A = \begin{pmatrix} a & c \\ b & d \end{pmatrix} \in \mathbf{R}^{2\times 2}.$$

The determinant of A is

$$\det A = ad - bc.$$

Let

$$J(u, v) = \det \begin{pmatrix} \frac{\partial h_1}{\partial u}(u, v) & \frac{\partial h_1}{\partial v}(u, v) \\ \\ \frac{\partial h_2}{\partial u}(u, v) & \frac{\partial h_2}{\partial v}(u, v) \end{pmatrix}.$$

This function is called the Jacobian of the vector function $h = (h_1, h_2)$; it is calculated as the determinant of a matrix. Therefore

$$J(u, v) = \frac{\partial h_1}{\partial u}(u, v)\frac{\partial h_2}{\partial v}(u, v) - \frac{\partial h_2}{\partial u}(u, v)\frac{\partial h_1}{\partial v}(u, v).$$

We assume that the derivatives here exist and that

$$J(u, v) \neq 0 \quad \forall x, y.$$

Theorem 4.17 *The joint density for U and V is*

$$f_{UV}(u, v) = f_{XY}(h_1(u, v), h_2(u, v))|J(u, v)|.$$

The proof of this theorem is based on the change of variables in double integrals; we will omit this proof.

Example 4.18 Let X and Y have joint density $f(x, y)$, $U = X$ and $V = X + Y$ (i.e., $g_1(x, y) = x$, $g_2(x, y) = x + y$). This gives

$$X = h_1(U, V) = U, \qquad Y = h_2(U, V) = V - U.$$

Hence, the Jacobian is

$$J(u,v) = \det \begin{pmatrix} \frac{\partial h_1}{\partial u}(u,v) & \frac{\partial h_1}{\partial v}(u,v) \\ \frac{\partial h_2}{\partial u}(u,v) & \frac{\partial h_2}{\partial v}(u,v) \end{pmatrix} = \det \begin{pmatrix} 1 & 0 \\ -1 & 1 \end{pmatrix} = 1 \cdot 1 + 1 \cdot 0 = 1.$$

Therefore, the joint density for U and V is

$$f_{UV}(u,v) = f_{XY}(u, v-u).$$

Corollary 4.19 *The example above gives us a way to derive the density for the sum $X+Y$ as the marginal density for $V = X+Y$ in the joint distribution (X,V), i.e.,*

$$f_V(v) = \int_{-\infty}^{\infty} f_{XY}(u, v-u)du.$$

Example 4.20 Assume that X and Y are independent and distributed uniformly in $[0,1]$. Let $U = X$ and $V = X+Y$. The joint density for U and V is

$$f_{UV}(u,v) = f_{XY}(u,v-u) = \mathbb{I}_{[0,1]\times[0,1]}(u,v-u) = \mathbb{I}_{[0,1]}(u)\mathbb{I}_{[0,1]}(v-u).$$

It gives the marginal density for $V = X+Y$ as

$$f_V(v) = \int_{-\infty}^{\infty} f_{XY}(u,v-u)du = \int_{-\infty}^{\infty} \mathbb{I}_{[0,1]}(u)\mathbb{I}_{[0,1]}(v-u)du$$
$$= \int_0^1 \mathbb{I}_{[0,1]}(v-u)du.$$

It can be seen that the domain where $\mathbb{I}_{[0,1]}(u)\mathbb{I}_{[0,1]}(v-u) \neq 0$ has a rhombus shape. To calculate the density $f_V(v)$ for V, it suffices to consider $v \in [0,2]$ only.

For $v \in [0,1]$,

$$v - u \in [0,1] \quad \text{iff} \quad u \in [0,v].$$

In this case,

$$f_V(v) = \int_0^1 \mathbb{I}_{[0,1]}(v-u)du = \int_0^v du = v.$$

For $v \in [1,2]$,

$$v - u \in [0,1] \quad \text{iff} \quad u \in [v-1,1].$$

Likewise,

$$f_V(v) = \int_0^1 \mathbb{I}_{[0,1]}(v-u)du = \int_{v-1}^1 du = 2 - v.$$

It can be verified immediately that $\int_0^2 f_V(v)dv = 1$.

4.5 Multivariate case

Theorem 4.17 can be extended on the transformations of more than two random variables.

Let $X = (X_1, ..., X_n)$ have a joint density $f_X(x) = f_X(x_1, ..., x_n)$ defined for $x \in \mathbf{R}^n$, let $g_i : \mathbf{R}^n \to \mathbf{R}$ be differentiable functions, and let the new random variables be

$$Y_i = g_i(X_1, ..., X_n).$$

We assume that the transformation can be inverted, i.e., that there exist functions $h_i : \mathbf{R}^n \to \mathbf{R}$ such that

$$X_i = h_i(Y_1, ..., Y_n).$$

We assume that the $J(y) \neq 0$ for all $y = (y_1, ..., y_n)$, where the Jacobian is

$$J(y) = \det \begin{pmatrix} \frac{\partial h_1}{\partial y_1}(y) \ ... \frac{\partial h_1}{\partial y_n}(y) \\ .... \qquad ... \\ \frac{\partial h_n}{\partial y_1}(y) \ ... \frac{\partial h_n}{\partial y_n}(y) \end{pmatrix}.$$

Here, det is the determinant of a matrix. In the examples and problems used in this course, we will not be calculating determinants for $n > 2$.

Theorem 4.21 *The joint density for* $Y = (Y_1, ..., Y_n)$ *is*

$$f_Y(y) = f_X(h_1(y), ..., h_n(y))|J(y))|,$$

where $y = (y_1, ..., y_n)$.

4.6 Distributions for maximums and minimums

Distribution for maximums

Assume that $X_1, X_2, ..., X_n$ are independent random variables with the c.d.f. F and its density f, and

$$U = \max(X_1, ..., X_n).$$

Let us find the density for U.

First, we note that

$$U \leq u \quad \text{iff} \quad X_i \leq u \quad \text{for all } i.$$

Thus

$$F_U(u) = \mathbf{P}(U \leq u) = \mathbf{P}(X_1 \leq u, X_2 \leq u, \ldots, X_n < u).$$

By the assumption about independence, we obtain

$$F_U(u) = \mathbf{P}(X_1 \le u)\mathbf{P}(X_2 \le u) \times \cdots \times \mathbf{P}(X_n \le u) = F(u)^n.$$

Differentiating, its density is

$$f_U(u) = \frac{d}{du}(F(u)^n) = nF(u)^{n-1}f(u).$$

Example 4.22 Let random variables X and Y be independent and uniformly distributed in $[0,1]$, with the c.d.f.

$$F(x) = \begin{cases} 1, & x > 1 \\ x, & x \in [0,1] \\ 0, & x < 0 \end{cases}$$

and the density $f(x) = \mathbb{I}_{[0,1]}(x)$. Let $U = \max(X, Y)$. The density of U is

$$f_U(u) = 2u\mathbb{I}_{[0,1]}(u).$$

It can be verified immediately that

$$\int_{-\infty}^{\infty} f_U(u)du = \int_0^1 2udu = \frac{2u^2}{2}\bigg|_0^1 = 1.$$

Example 4.23 Let X_1, X_2, and X_3, be independent and uniformly distributed in $[0,1]$ and

$$U = \max(X_1, X_2, X_3).$$

The density of U is

$$f_U(u) = 3u^2\mathbb{I}_{[0,1]}(u).$$

Again, it can be verified immediately that

$$\int_{-\infty}^{\infty} f_U(u)du = \int_0^1 3u^2du = \frac{3u^3}{3}\bigg|_0^1 = 1.$$

Distribution for minimums

Let us assume that $X_1, X_2, ..., X_n$ are independent random variables with the c.d.f. F and density f, and

$$V = \min(X_1, ..., X_n).$$

Let us find the density for V.

We note that

$$V > v \quad \text{iff} \quad X_i > v \quad \text{for all } i.$$

Thus

$$F_V(v) = 1 - \mathbf{P}(V > v) = 1 - \mathbf{P}(X_1 > v, X_2 > v, ..., X_n > v).$$

By the assumption about independence, we obtain

$$F_V(v) = 1 - \mathbf{P}(X_1 > v)\mathbf{P}(X_2 > v) \times ... \times \mathbf{P}(X_n > v) = 1 - (1 - F(v))^n.$$

Differentiating, its density is

$$f_V(v) = \frac{d}{dv}(1 - (1 - F(v))^n) = n(1 - F(v))^{n-1} f(v).$$

Example 4.24 Let X and Y be independent and uniformly distributed in $[0, 1]$ and $V = \min(X, Y)$.

The density of V is

$$f_V(v) = 2(1 - v)\mathbb{I}_{[0,1]}(v).$$

Example 4.25 Let X_1, X_2, X_3 be independent and uniformly distributed in $[0, 1]$ and

$$V = \min(X_1, X_2, X_3).$$

The density of V is

$$f_V(v) = 3(1 - v)^2\mathbb{I}_{[0,1]}(v).$$

The examples above provide us with a method of creating random variables with certain given densities. For instance, a random variable with linear growth of density on $[0, 1]$ can be created as $\max(X_1, X_2)$. A random variable with quadratic growth of density on $[0, 1]$ can be created as $\max(X_1, X_2, X_2)$, where X_i. The random variables here X_i have to be generated by a standard random number generator that gives uniform distribution on $[0, 1]$ such as *rand* in MATLAB.

4.7 Order statistics

This section is concerned with ordering a collection of independent continuous random variables.

Definition 4.26 *Consider random variables $X_1, X_2, ..., X_n$. We call the kth-smallest value kth order statistic $X_{(k)}$. In other words, we sort values and denote by $X_{(1)} \leq X_{(2)} \leq \cdots \leq X_{(n)}$. We call these new random variables* order statistics.

In particular,

$$X_{(1)} = \min(X_1, ..., X_n), \quad X_{(n)} = \max(X_1, ..., X_n).$$

Let us assume that $X_1, X_2, ..., X_n$ are independent random variables with the c.d.f. F and density f. Let us find the density for their order statistics.

Let $1 < k < n$. The event $x < X_{(k)} \leq x + dx$ occurs if: (a) $k-1$ observations are less than x; (b) one observation is in the interval $[x, x+dx]$; and (c) $n-k$ observations are greater than $x + dx$.

The probability of any particular arrangement of this type is

$$f(x)F(x)^{k-1}(1-F(x))^{n-k}dx.$$

By the combinatorial rule from Week 1, there are

$$\frac{n!}{(k-1)!1!(n-k)!}$$

such arrangements. It follows that the probability that at least one of these arrangements occurs is

$$\frac{n!}{(k-1)!(n-k)!}f(x)F(x)^{k-1}(1-F(x))^{n-k}dx.$$

Hence, the density for kth order statistics is

$$f_k(x) = \frac{n!}{(k-1)!(n-k)!}f(x)F(x)^{k-1}(1-F(x))^{n-k}.$$

Example 4.27 Let X_i be independent and uniformly distributed in $[0, 1]$. The density of $X_{(k)}$ is

$$f_k(x) = \frac{n!}{(k-1)!(n-k)!}x^{k-1}(1-x)^{n-k}\mathbb{I}_{[0,1]}(x).$$

Example 4.28 Consider a model where the future profits of three companies are random, independent, and distributed uniformly in the interval [−0.2M, 1M]. Find the probability that the profit will be positive for the company in the middle (i.e., the second with respect to the profit). We consider random variables X_i that are independent and uniformly distributed in $[-0.2, 1]$.

Solution. The density for X_i is

$$f(x) = \frac{1}{1.2}\mathbb{I}_{[-0.2,1]}(x),$$

and the c.d.f. is

$$F(x) = \begin{cases} 1, & x > 1 \\ \frac{1}{1.2}(x+0.2), & x \in [-0.2, 1] \\ 0, & x < -0.2. \end{cases}$$

We need to find

$$\mathbf{P}(X_{(2)} > 0) = \int_0^1 f_2(x)dx,$$

where f_2 is the density of 2nd order statistics, with

$$f_2(x) = \frac{3!}{1!1!}\frac{1}{1.2}(x+0.2)\left(1 - \frac{1}{1.2}(x+0.2)\right)\frac{1}{1.2}\mathbb{I}_{[-0.2,1]}(x).$$

Integrating, we obtain that $\int_0^1 f_2(x)dx = 25/27$.

Problems for Week 4

Problem 4.1 *Let X have exponential distribution $X \sim Exp(\lambda)$. Find the density for $U = -X/2$.*

Problem 4.2 *Let X have exponential distribution $X \sim Exp(\lambda)$. Find the density for $U = -X/2 + 3$.*

Problem 4.3 *Let X have the density $f_X(x)$ and $Y = X^2$. Find the c.d.f. $F_Y(y)$ and the density $f_Y(y)$ for $Y = X^2$.*

Problem 4.4 *Let X and Y have exponential distributions $X \sim Exp(\alpha)$ and $Y \sim Exp(\beta)$ We assume that X and Y are independent. Let $U = X + 2Y$ and $V = X + Y$.*

(a) Find the joint density for (U, V).
(b) Find an integral formula for the marginal density of V.
(c) Find explicitly the marginal density of V for the case where $\alpha \neq \beta$.
(d) Find explicitly the marginal density of V for the case where $\alpha = \beta$.

Problem 4.5 *Find the density (if exists) for $U = \max(X, Y)$, where $X \sim U(0, 1)$ (i.e. a uniform distribution), $Y \sim Exp(\alpha)$ (i.e. an exponential distribution). We assume that X and Y are independent.*

Week 5: Expectation of Random Variables

In this chapter, we introduce expectations of random variables and discuss their calculation.

5.1 Expectation of a discrete random variable

We introduced previously a median as a characteristic of a "center" of the distribution. There is an alternative and more popular parameter of distribution that describes its center.

Definition 5.1 *Let X be a discrete random variable with the probability frequency function $p(x)$ such that*

$$\sum_x |x|p(x) < +\infty.$$

The expectation *of X, or the expected value of X, or the* mean value *of X, is*

$$\mathbf{E}(X) = \sum_x xp(x).$$

Note that the condition $\sum_x |x|p(x) < +\infty$ is always satisfied for the random variable taking values at finite sets.
Notation: It is acceptable to write $\mathbf{E}X$ *instead of* $\mathbf{E}(X)$.

Example 5.2 Let X be a Bernoulli random variable such that

$$X = \begin{cases} 1, & \text{with probability } p \\ 0, & \text{with probability } 1-p. \end{cases}$$

The expectation of X is $\mathbf{E}X = 1 \cdot p + 0 \cdot (1-p) = p$.

Note that X takes values 0 or 1 only and it never takes the value $p = \mathbf{E}X$, if $p \in (0, 1)$. This is why the term "the expected value" should not be taken literally.

Example 5.3 A gambler bets on the coin tossing game: he pays \$100 for head and receives \$100 for the tail. Let us find the expected gain.

Let X be the gain. It is a random variable with frequency $p_X(x)$ given by

x	-100	100
$p_X(x)$	$1/2$	$1/2$

It gives $\mathbf{E}X = 100(1/2) + 100(-1/2) = 0$. Therefore, this is a fair game.

Example 5.4 A roulette wheel has the numbers 1, 2, ..., 36, as well as 0 and 00 (total 38 numbers). If you bet \$1 that an odd non-zero number comes up, you win or lose \$1 according to whether or not that event occurs. Let X be the gain. We have

$$\mathbf{P}(X = -1) = 20/38, \quad \mathbf{P}(X = 1) = 18/38.$$

Hence

$$\mathbf{E}X = -1 \cdot 20/38 + 1 \cdot 18/38 = -1/19.$$

Your expected loss is about \$0.05. This game gives an advantage to the casino; it covers the casino business expenses.

Example 5.5 Consider the roulette wheel again. If you bet \$1 that the number 1 comes up, you win \$35 or lose \$1 according to whether or not that event occurs. Let X be the gain. We have

$$\mathbf{P}(X = -1) = 37/38, \quad \mathbf{P}(X = 35) = 1/38.$$

Hence

$$\mathbf{E}X = -1 \cdot 37/38 + 35 \cdot 1/38 = -1/19.$$

Your expected loss is about \$0.05 again. Note that the expected value is the same as for the previous example.

Theorem 5.6 *Let a be a constant. The expected value of a is*

$$\mathbf{E}(a) = a.$$

Proof. a can be described as a discrete random variable that takes only one value, a. Then $\mathbf{E}a = a \cdot 1$.

5.2 Expectation of a continuous random variable

Definition 5.7 *Let X be a continuous random variable with the probability density function $f(x)$ such that*

$$\int_{-\infty}^{\infty} |x| f(x) dx < +\infty.$$

The expectation of X, or the expected value of X, or the mean value of X, is

$$\mathbf{E}X = \mathbf{E}(X) = \int_{-\infty}^{\infty} x f(x) dx.$$

Example 5.8 Let X have uniform distribution $U(0,a)$, i.e., it has the density

$$f(x) = \begin{cases} \frac{1}{a}, & 0 \leq x \leq a \\ 0, & \text{elsewhere.} \end{cases}$$

The expected value of X is

$$\mathbf{E}X = \int_0^a \frac{x}{a} dx = \frac{1}{a} \frac{x^2}{2} \bigg|_0^a = \frac{a}{2}.$$

Example 5.9 Let X have the probability density function

$$f(x) = \begin{cases} \lambda e^{-\lambda x}, \ x \geq 0 \\ 0, \qquad \text{elsewhere.} \end{cases}$$

The expected value of X is

$$\mathbf{E}X = \mathbf{E}(X) = \int_0^{+\infty} x \lambda e^{-\lambda x} dx = \frac{1}{\lambda}.$$

5.3 The case of the distributions with heavy tails (fat tails)

It may happen that

$$\int_{-\infty}^{\infty} |x| f(x) dx = +\infty$$

(i.e., the integral diverges). It happens if $f(x)$ is not decaying fast enough when $|x| \to +\infty$ (the case of heavy (fat) tail distribution). Technically, the integral $\int_{-\infty}^{\infty} x f(x) dx$ is not defined in this case.

It is common to accept the following rules:

(A) If

$$\int_{-\infty}^{0} xf(x)dx > -\infty, \quad \int_{0}^{\infty} xf(x)dx = +\infty$$

then $\mathbf{E}X = +\infty$ (and, therefore, it is defined).

(B) If

$$\int_{-\infty}^{0} xf(x)dx = -\infty, \quad \int_{0}^{\infty} xf(x)dx < +\infty$$

then $\mathbf{E}X = -\infty$ (and, therefore, it is defined).

The only case where $\mathbf{E}X$ is not defined in any sense is the case where

$$\int_{-\infty}^{0} xf(x)dx = -\infty, \quad \int_{0}^{\infty} xf(x)dx = +\infty. \tag{5.3}$$

Similar rules are accepted for the case of discrete random variables.

Example 5.10 Consider a random variable $X = X_1/X_2$, where $X_i \sim N(0,1)$. We had found in Week 4 that X has the density

$$f(x) = \frac{1}{\pi(1+x^2)}$$

(i.e., a Cauchy density). For this density, (5.3) holds, since have that

$$\int_{-\infty}^{0} \frac{x}{\pi(1+x^2)}dx = -\infty, \quad \int_{0}^{\infty} x\frac{x}{\pi(1+x^2)}dx = +\infty.$$

This means that $\mathbf{E}X$ is not defined.

5.4 Expectations of functions of random variables

Discrete random variables

Theorem 5.11 *Let $g(x)$ be a function, $g : \mathbf{R} \to \mathbf{R}$ and X be a discrete random variable with frequency $p(x)$ such that*

$$\sum_x |g(x)|p(x) < +\infty.$$

For $Y = g(X)$,

$$\mathbf{E}Y = \sum_x g(x)p(x).$$

This theorem shows that we can calculate $\mathbf{E}g(X)$ without using the probability frequency function of $Y = g(X)$.

Proof. Assume that X take values x_k, $k = 1, 2, \ldots$, and $\mathbf{P}(X = x_k) = p_X(x_k)$. It follows that Y take values $y_k = g(x_k)$, $k = 1, 2, \ldots$. Assume first that, for any y_k, there is a unique x_k such that $y_k = g(x_k)$. Then $\mathbf{P}(Y = y_k) = p_X(x_k)$. It is followed that the frequency function of Y is $p_Y(y_k) = p_X(x_k)$. Hence

$$\mathbf{E}Y = \sum_k y_k p_Y(y_k) = \sum_k g(x_k) p_X(x_k) = \sum_x g(x) p(x).$$

For the case when there are more than one x such that $y_k = g(x)$ requires additional analysis, we have

$$\mathbf{P}(Y = y_k) = \sum_{j:\ g(x_j) = y_k} p_X(x_j).$$

Hence

$$\begin{aligned}\mathbf{E}Y &= \sum_k y_k p_Y(y_k) = \sum_k y_k \sum_{j:\ g(x_j) = y_k} p_X(x_j) \\ &= \sum_k \sum_{j:\ g(x_j) = y_k} g(x_j) p_X(x_j) = \sum_x g(x) p(x).\end{aligned}$$

Example 5.12 Let X be a Bernoulli random variable such that

$$X = \begin{cases} 1, & \text{with probability } p \\ 0, & \text{with probability } 1 - p. \end{cases}$$

We have that $\mathbf{E}(X^2) = 1^2 \cdot p + 0^2 \cdot (1 - p) = p$.

Notations: It is common to write $\mathbf{E}X^2$ instead of $\mathbf{E}(X^2)$. Therefore, $\mathbf{E}X^2$ means $\mathbf{E}(X^2)$ rather than $(\mathbf{E}X)^2$.

The case of continuous random variables

Let $g(x)$ be a function, $g : \mathbf{R} \to \mathbf{R}$ and X be a continuous random variable with density $f(x)$ such that

$$\int_{-\infty}^{\infty} |g(x)| f(x) dx < +\infty.$$

The expectation of $Y = g(X)$ is

$$\mathbf{E}Y = \int_{-\infty}^{\infty} g(x) f(x) dx.$$

Example 5.13 Consider the uniform distribution $U(0, a)$ with the density

$$f(x) = \begin{cases} \dfrac{1}{a}, & 0 \leq x \leq a \\ 0, & \text{elsewhere.} \end{cases}$$

Hence

$$\mathbf{E}(X^2) = \int_0^a \frac{x^2}{a} dx = \frac{1}{a}\frac{x^3}{3}\bigg|_0^a = \frac{a^2}{3}.$$

Example 5.14 Consider a distribution with the density

$$f(x) = \begin{cases} \lambda e^{-\lambda x}, & x \geq 0 \\ 0, & \text{elsewhere.} \end{cases}$$

Thus

$$\mathbf{E}X^2 = \mathbf{E}(X^2) = \int_0^{+\infty} x^2 \lambda e^{-\lambda x} dx = \frac{2}{\lambda^2}.$$

5.5 Joint probability distributions

Theorem 5.15 *Let X and Y be jointly distributed discrete random variables with the probability distribution function $f(x, y)$. Then*

$$\mathbf{E}g(X, Y) = \sum_x \sum_y g(x, y) f(x, y).$$

Theorem 5.16 *Let X and Y be jointly distributed with the probability distribution function $f(x, y)$. Then*

$$\mathbf{E}g(X, Y) = \int_{-\infty}^{\infty} \int_{-\infty}^{\infty} g(x, y) f(x, y) dx dy.$$

5.6 Expectation of linear combination of random variables

Theorem 5.17 *Let X be a random variable and a constant $a \in \mathbf{R}$,*

$$\mathbf{E}(aX) = a\mathbf{E}X.$$

Proof. We consider the continuous case only. Take $g(x) = ax$. We have

$$\mathbf{E}(aX) = \int_{-\infty}^{\infty} axf(x)dx = a \int_{-\infty}^{\infty} xf(x)dx = a\mathbf{E}X.$$

Example 5.18 A roulette wheel has the numbers 1, 2, .., 36, as well as 0 and 00 (total 38 numbers). If you bet \$1 that an odd non-zero number comes up, you win or lose \$1 according to whether or not that event occurs. Let X be the gain. We found above that

$$\mathbf{E}X = -\$1/19 \approx -\$0.05.$$

If you bet \$100 that an odd non-zero number comes up, you win or lose $Y = 100X$. Respectively, $\mathbf{E}Y = 100\mathbf{E}X \approx -\$5$.

Theorem 5.19 *Let X be a random variable and $a \in \mathbf{R}$,*

$$\mathbf{E}(a + X) = a + \mathbf{E}X.$$

Proof. Again, we consider the continuous case only. Take $g(x) = a + x$. We have

$$\mathbf{E}(a + X) = \int_{-\infty}^{\infty} (a + x)f(x)dx = a\int_{-\infty}^{\infty} f(x)dx + \int_{-\infty}^{\infty} xf(x)dx = a + \mathbf{E}X.$$

Example 5.20 A roulette wheel has the numbers 1, 2, ..., 36, as well as 0 and 00 (total 38 numbers). If you bet \$1 that an odd non-zero number comes up, you win or lose \$1 according to whether or not that event occurs. Let X be the gain. We found above that $\mathbf{E}X = -\$1/19 \sim -\$0.05$. Assume that you have \$10 in your pocket, and let Y be the total amount of money in your pocket after the game. Then $Y = \$10 + X$ and $\mathbf{E}Y = \$10 + \mathbf{E}X \sim \$9.95$.

Theorem 5.21 *Let X and Y be random variables,*

$$\mathbf{E}(X + Y) = \mathbf{E}X + \mathbf{E}Y.$$

Proof. We consider continuous case only. Take $g(x, y) = x + y$. We have:

$$\begin{aligned}
\mathbf{E}(X + Y) &= \int_{-\infty}^{\infty}\int_{-\infty}^{\infty} (x + y)f(x, y)dxdy \\
&= \int_{-\infty}^{\infty} x\left\{\int_{-\infty}^{\infty} f(x, y)dy\right\} dx + \int_{-\infty}^{\infty} y\left\{\int_{-\infty}^{\infty} f(x, y)dx\right\} \\
&= \int_{-\infty}^{\infty} xf_X(x)dx + \int_{-\infty}^{\infty} yf_Y(y)dy \\
&= \mathbf{E}X + \mathbf{E}Y.
\end{aligned}$$

Here, $f_X(x)$ and $f_Y(y)$ are the marginal distributions of X and Y, respectively.

Example 5.22 A roulette wheel has the numbers 1, 2, ..., 36, as well as 0 and 00 (total 38 numbers). If you bet \$1 that an odd non-zero number comes up, you win or lose \$1 according to whether or not that event occurs. Let X be the gain. If you bet another \$1 that the number 1 comes up, you win \$35 or lose \$1 according to whether or not that event occurs. Let Y be the gain. We found above that $\mathbf{E}X = \mathbf{E}Y = -\$1/19 \approx -\$0.05$. The expected gain for this combined \$2 bet is $\mathbf{E}(X+Y) = -2/19 \approx -\$0.1$.

Corollary 5.23 *Let $X_1, ..., X_n$ be random variables, $a_i \in \mathbf{R}$, and*

$$Z = a_0 + \sum_{i=1}^{n} a_i X_i.$$

Then

$$\mathbf{E}Z = a_0 + \sum_{i=1}^{n} a_i \mathbf{E}X_i.$$

Example 5.24 Let X be the total number of successes in n Bernoulli trials, i.e. $X = X_1 + ... + X_n$, where X_i are Bernoulli variables such that

$$X_i = \begin{cases} 1, & \text{with probability } p \\ 0, & \text{with probability } 1-p. \end{cases}$$

For each X_i, $\mathbf{E}X_i = 1 \cdot p + 0 \cdot (1-p) = p$. Hence, $\mathbf{E}X = p + ... + p = np$.

Corollary 5.25 *Let $S(X)$ and $T(X)$ be functions of X. Then*

$$\mathbf{E}(S(X) + T(X)) = \mathbf{E}(S(X)) + \mathbf{E}(T(X)).$$

Corollary 5.26 *Let X and Y be jointly distributed with g and h be functions of the random variables X and Y, respectively. Then*

$$\mathbf{E}[g(X,Y)] + h(X,Y)] = \mathbf{E}g(X,Y) + \mathbf{E}h(X,Y).$$

5.7 Expectation of a product

Let us consider the following question: *Is it correct that* $\mathbf{E}(XY) = \mathbf{E}(X)\mathbf{E}(Y)$? The answer is no. For example, take random variable X such that

$$X = \begin{cases} 1, & \text{with probability } 1/2 \\ -1, & \text{with probability } 1/2. \end{cases}$$

We have that $\mathbf{E}X = 0$, but $X^2 = 1$ with probability 1 and $\mathbf{E}X^2 = 1 \neq \mathbf{E}X \cdot \mathbf{E}X = 0$.

Theorem 5.27 *If X and Y are independent random variables, then*

$$\mathbf{E}(XY) = \mathbf{E}(X)\mathbf{E}(Y).$$

Proof. Consider the continuous case only. We have

$$\mathbf{E}(XY) = \int_{-\infty}^{\infty}\int_{-\infty}^{\infty} xyf(x,y)dxdy.$$

Since X and Y are independent, we have $f(x,y) = f_X(x)f_Y(y)$, where $f_X(x)$ and $f_Y(y)$ are the marginal distributions of X and Y, respectively. Hence

$$\begin{aligned}\mathbf{E}(XY) &= \int_{-\infty}^{\infty}\int_{-\infty}^{\infty} xyf(x,y)dxdy = \int_{-\infty}^{\infty}\int_{-\infty}^{\infty} xyf_X(x)f_Y(y)dxdy \\ &= \int_{-\infty}^{\infty} xf_X(x)dx \int_{-\infty}^{\infty} yf_Y(y)dy = \mathbf{E}(X)\mathbf{E}(Y).\end{aligned}$$

It can be noted that the condition $\mathbf{E}(XY) = \mathbf{E}(X)\mathbf{E}(Y)$ is necessary but not sufficient for independency of X and Y. In other words, $\mathbf{E}(XY) = \mathbf{E}(X)\mathbf{E}(Y)$ does not necessarily imply that X and Y are independent.

5.8 Probability of an event

Let A be a random event. Consider a random variable $\mathbb{I}_A$ such that

$$\mathbb{I}_A = \begin{cases} 1, & \text{if event } A \text{ occurs} \\ 0, & \text{otherwise.} \end{cases}$$

Theorem 5.28 $\mathbf{E}\mathbb{I}_A = \mathbf{P}(A)$.

Expectation in the axiomatic setting

Let Ω be the sample space, and let $X(\omega)$ be a random variable, i.e., it is a mapping $X : \Omega \to \mathbf{R}$. In this setting, it is common to write

$$\mathbf{E}X = \int_{\Omega} X(\omega)\mathbf{P}(d\omega),$$

meaning that the expectation is a kind of an integral. More advanced probability theory courses gives an interpretation to this integration.

5.9 The moments

Definition 5.29 *The kth moment of the random variable is defined as*

$$\mu_k^{'} = \mathbf{E}(X^k).$$

Those moments are usually denoted by $\mu_k^{'}$.

The first moment (i.e., $\mathbf{E}X$) is called the *mean value* (mean) of X and is often denoted by μ, i.e.,

$$\mu_1^{'} = \mathbf{E}(X) = \text{ mean of } X \equiv \mu.$$

Example 5.30 Let X be a Bernoulli random variable such that

$$X = \begin{cases} 1, & \text{with probability } p \\ 0, & \text{with probability } 1-p. \end{cases}$$

We have that $\mu_k' = \mathbf{E}(X^k) = 1^k \cdot p + 0^k \cdot (1-p) = p$.

Definition 5.31 *The kth central moment (or the moment about the mean) of the random variable X is defined as* $\mathbf{E}[(X-\mu)^k]$.

The kth central moment is usually denoted by μ_k. By the rule of expectation, we have that

$$\mu_k = \begin{cases} \sum_x (x-\mu)^k f(x) & \text{if } X \text{ is discrete} \\ \int_{-\infty}^{\infty} (x-\mu)^k f(x) dx & \text{if } X \text{ is continuous.} \end{cases}$$

Example 5.32 Let X be a Bernoulli random variable such that

$$X = \begin{cases} 1, & \text{with probability } p \\ 0, & \text{with probability } 1-p. \end{cases}$$

We have that $\mu = \mathbf{E}X = p$,

$$\mu_k = \mathbf{E}(X-\mu)^k = (1-p)^k \cdot p + (-p)^k \cdot (1-p).$$

In particular,

$$\begin{aligned} \mu_2 = (1-p)^2 \cdot p + (-p)^2 \cdot (1-p) &= p - 2p^2 + p^3 - p^2 - p^3 \\ &= p - p^2 = p(1-p). \end{aligned}$$

Problems for Week 5

Problem 5.1 *Assume that the probability frequency function for a random variable X is defined as*

x	*1*	*3*	*4*	*5*
$p(x)$	*0.1*	*0.2*	*0.5*	*0.2*

Find $\mathbf{E}X$ *and* $\operatorname{Var}(X)$.

Problem 5.2 *Assume that the probability frequency function for a random variable X is defined as*

x	*1*	*3*	*4*	*5*
$p(x)$	*0.2*	*0.3*	*0.2*	k

Here $k \in \mathbf{R}$ is some unknown value. Find $\mathbf{E}(X)$.

Problem 5.3 *Let a random variable X have the probability density function $f(x)$ such that $f(x) = x/8$ if $0 < x < 4$, and $f(x) = 0$ otherwise. Find* $\mathbf{E}X$ *and* $\operatorname{Var} X$.

Problem 5.4 *Let the probability density function $f(x)$ of a random variable X be given in Fig. 5.1: $f(x) = 0$ for $x < 2$ or $x > 8$, $f(x)$ is linear for $x \in [2, 8]$, $f(8) = h$.*

(a) Find the value of h. (Hint: use the fact that the integral $\int_{-\infty}^{\infty} f(x)dx$ is known.)
(b) Find $\mathbf{E}X$ *and* $\operatorname{Var} X$.
(c) Find $\mathbf{P}(X < 0)$, $\mathbf{P}(X = 8)$, $\mathbf{P}(X < 5)$.
(d) Find the cumulative distribution function $F(x) = \mathbf{P}(X \leq x)$. Draw a sketch for F.

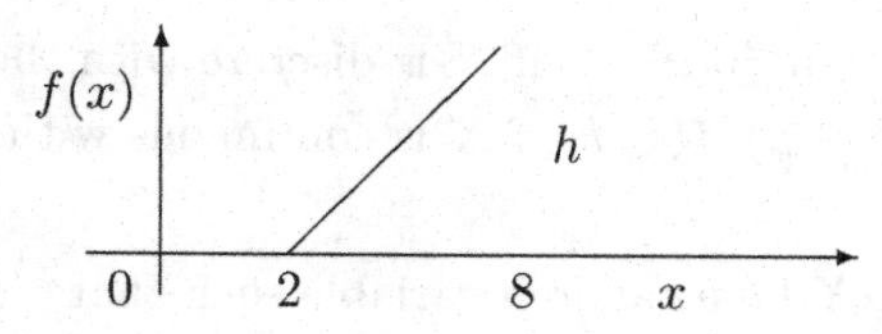

Figure 5.1: The shape of probability density function $f(x)$; h is unknown.

Problem 5.5 *Prove that $\mathbf{E}X = \lambda$ for $X \sim Poi(\lambda)$ (i.e., the Poisson distribution).*

Week 6. Variance and Covariance

In this chapter, we discuss variance, covariance and correlations of random variables.

6.1 Variance and standard deviation

Definition 6.1 *The variance of a random variable* X *is* $\operatorname{Var} X = \mathbf{E}(X - \mathbf{E}X)^2$.

It is the most common indication of how dispersed the distribution is about its center; for instance, it is commonly used in economics and finance as a measure of the risk associated with future uncertainty. In our notations, $\operatorname{Var} X = \mu_2$, i.e., it is the central 2nd moment.

Definition 6.2 *The standard deviation of a random variable* X *is* $\sigma = \sqrt{\operatorname{Var} X}$.

The standard deviation represents the average distance from the mean $\mu = \mathbf{E}X$, with

$$\operatorname{Var} X = \begin{cases} \sum_x (x-\mu)^2 p(x) & \text{if } X \text{ is discrete with the frequency } p, \\ \int_{-\infty}^{\infty} (x-\mu)^2 f(x) dx & \text{if } X \text{ is continuous with the density } f. \end{cases}$$

Example 6.3 Let X be a random variable such that

$$X = \begin{cases} 1, & \text{with probability } 1/2 \\ -1, & \text{with probability } 1/2. \end{cases}$$

It can be found that

$$\mathbf{E}X = 1 \cdot 1/2 + (-1) \cdot 1/2 = 0,$$

and

$$\operatorname{Var} X = 1^2 \cdot 1/2 + (-1)^2 \cdot 1/2 = 1.$$

Therefore, the standard deviation of X is $\sigma_X = 1$.

Example 6.4 Let Y be a random variable such that

$$Y = \begin{cases} 10, & \text{with probability } 1/2 \\ -10, & \text{with probability } 1/2. \end{cases}$$

We have

$$\mathbf{E}Y = 10 \cdot 1/2 + (-10) \cdot 1/2 = 0,$$

and

$$\operatorname{Var} Y = (10)^2 1^2 \cdot 1/2 + (-10)^2 \cdot 1/2 = 100.$$

Therefore, the standard deviation of Y is $\sigma_Y = 10$.

Example 6.5 Let X be a Bernoulli random variable such that

$$X = \begin{cases} 1, & \text{with probability } p \\ 0, & \text{with probability } 1-p. \end{cases}$$

The expectation and variance of X are

$$\mathbf{E}X = 1 \cdot p + 0 \cdot (1-p) = p,$$

$$\operatorname{Var} X = \mu_2 = (1-p)^2 \cdot p + (-p)^2 \cdot (1-p) = p - p^2.$$

Note that $\operatorname{Var} X = p - p^2$ achieves its maximum at $p = 1/2$. In this sense, the Bernoulli random variable is "the most random" if $p = 1/2$.

Theorem 6.6 $\operatorname{Var} X = \mathbf{E}(X^2) - (\mathbf{E}X)^2$. *(Assuming that the corresponding expectations exist.)*

Proof. Let $\mu = \mathbf{E}X$. We have

$$\mathbf{E}(X-\mu)^2 = \mathbf{E}[X^2 - 2X\mu + \mu^2] = \mathbf{E}(X^2) - 2\mathbf{E}X\mu + \mathbf{E}\mu^2 = \mathbf{E}(X^2) - \mu^2.$$

In some cases, this theorem helps to simplify calculations.

Example 6.7 Consider the uniform distribution over $[0, a]$ with the density

$$f(x) = \begin{cases} \frac{1}{a}, & 0 \le x \le a \\ 0, & \text{elsewhere.} \end{cases}$$

We found that $\mathbf{E}X = a/2$. Hence

$$\operatorname{Var} X = \int_0^a \left(x - \frac{a}{2}\right)^2 \frac{1}{a} dx = \frac{1}{a}\left(\frac{x^3}{3} - 2\frac{x^2 a}{4} + \frac{x a^2}{4}\right)\Big|_0^a = \frac{a^2}{12}.$$

Example 6.8 Let us apply Theorem 6.6. We have

$$\mathbf{E}X = a/2,$$

$$\mathbf{E}(X^2) = \int_0^a \frac{x^2}{a} dx = \frac{1}{a}\frac{x^3}{3}\bigg|_0^a = \frac{a^2}{3}.$$

Hence

$$\operatorname{Var} X = \frac{a^2}{3} - \frac{a^2}{4} = \frac{a^2}{12}.$$

Example 6.9 Let us apply Theorem 6.6 for a Bernoulli random variable X such that

$$X = \begin{cases} 1, & \text{with probability } p \\ 0, & \text{with probability } 1 - p. \end{cases}$$

We have $\mathbf{E}X = 1 \cdot p + 0 \cdot (1-p) = p$ and $\mathbf{E}(X^2) = 1^2 \cdot p + 0^2 \cdot (1-p) = p$. Hence $\operatorname{Var} X = p - p^2$.

6.2 Existence of variance and 2nd moment

Definition 6.10 *If* $\mathbf{E}(X^2) < +\infty$, *we say that* X *has a finite 2nd moment.*

Theorem 6.11 *If* $\mathbf{E}(X^2) < +\infty$, *then* $\operatorname{Var} X$ *is defined and* $\operatorname{Var} X < +\infty$.

Proof. It suffices to prove that $\mathbf{E}(X)$ is defined and $|\mathbf{E}(X)| < +\infty$. Let us consider continuous case. By the assumptions,

$$\int_{-\infty}^{\infty} x^2 f(x) dx < +\infty.$$

We have

$$\int_{-\infty}^{\infty} |x| f(x) dx = \int_{-\infty}^{-1} |x| f(x) dx + \int_{-1}^{1} |x| f(x) dx + \int_{1}^{\infty} |x| f(x) dx.$$

Hence

$$\int_{-1}^{1} |x| f(x) dx \leq \int_{-1}^{1} f(x) dx \leq 1.$$

It follows that

$$\int_{-\infty}^{-1} |x| f(x) dx + \int_{1}^{\infty} |x| f(x) dx \leq \int_{-\infty}^{-1} |x|^2 f(x) dx + \int_{1}^{\infty} |x|^2 f(x) dx$$
$$\leq \int_{-\infty}^{\infty} x^2 f(x) dx < +\infty.$$

6.3 Variance of linear transformations of random variables

Theorem 6.12 *Let X be a random variable, $a \in \mathbf{R}$, and $Y = a + X$. Then*

$$\operatorname{Var} Y = \operatorname{Var} X.$$

Proof.

$$\begin{aligned}\operatorname{Var} Y &= \mathbf{E}(Y - \mathbf{E}Y)^2 = \mathbf{E}(a + X - \mathbf{E}(a + X))^2 \\ &= \mathbf{E}(a + X - a - \mathbf{E}X))^2 = \mathbf{E}(X - \mathbf{E}X))^2 = \operatorname{Var} X.\end{aligned}$$

Example 6.13 Let $p \in [0, 1]$ be given and Y be a random variable such that

$$Y = \begin{cases} 99, & \text{with probability } p \\ 98, & \text{with probability } 1 - p. \end{cases}$$

Assume $Y = 98 + X$, where X is the Bernoulli random variable. We found that $\operatorname{Var} X = p - p^2$. Hence $\operatorname{Var} Y = p - p^2$.

Theorem 6.14 *Let X be a random variable, $a \in \mathbf{R}$, and $Y = a + bX$. Then* $\operatorname{Var} Y = b^2 \operatorname{Var} X$.

Proof.

$$\begin{aligned}\operatorname{Var} Y &= \mathbf{E}(Y - \mathbf{E}Y)^2 = \mathbf{E}(a + bX - \mathbf{E}(a + bX))^2 \\ &= \mathbf{E}(a + bX - a - b\mathbf{E}X))^2 = \mathbf{E}b^2(X - \mathbf{E}X))^2 = b^2 \operatorname{Var} X.\end{aligned}$$

Example 6.15 Let $p \in [0, 1]$ be given and X be a random variable such that

$$Y = \begin{cases} 99, & \text{with probability } p \\ 88, & \text{with probability } 1 - p. \end{cases}$$

Assume $Y = 88 + 11 \cdot X$, where X is the Bernoulli random variable. We found that $\operatorname{Var} X = p - p^2$. Hence $\operatorname{Var} Y = 121 \cdot (p - p^2)$.

6.4 Chebyshev's inequality

Theorem 6.16 *Let X be a random variable such that $\mathbf{E}(X^2) < +\infty$, $\mathbf{E}X = \mu$ and $\operatorname{Var} X = \sigma^2 > 0$. Then for any $t > 0$,*

$$\mathbf{P}(|X - \mu| > t) \leq \frac{\sigma^2}{t^2}.$$

Proof. We will provide the proof for the case where X is a continuous random variable. Let $D = \{x : |x - \mu| > t\}$. We have

$$P(|X - \mu| > t) = \int_D f(x)dx \leq \int_D \frac{(x-\mu)^2}{t^2} f(x)dx \leq \frac{\sigma^2}{t^2}.$$

This completes the proof.

Example 6.17 If $t = 4\sigma$, then the Chebyshev's inequality gives

$$\mathbf{P}(|X - \mu| > 4\sigma) \leq \frac{\sigma^2}{16\sigma^2} = \frac{1}{16}.$$

Corollary 6.18 *If* $\mathrm{Var}\, X = 0$ *then* $\mathbf{P}(X = \mu) = 1$, *i.e.,* X *is nonrandom.*

Proof. If $\mathbf{P}(X = \mu) < 1$, then there exists $\varepsilon > 0$ such that $\mathbf{P}(|X - \mu| > \varepsilon) > 0$. However, by the Chebyshev's inequality,

$$\mathbf{P}(|X - \mu| > \varepsilon) \leq \sigma^2/\varepsilon = 0/\varepsilon = 0.$$

Example 6.19 If $t = 2\sigma$, we obtain

$$\mathbf{P}(|X - \mu| > 2\sigma) \leq \frac{\sigma^2}{4\sigma^2} = \frac{1}{4}.$$

If $t = \sigma$, then the Chebyshev's inequality gives

$$\mathbf{P}(|X - \mu| > \sigma) \leq \frac{\sigma^2}{\sigma^2} = 1.$$

Therefore, the Chebyshev's inequality does not give any useful information in this case.

In the general case, the Chebyshev's inequality is not the "best" estimate. For some particular distributions, it can be improved.

Example 6.20 Let X be uniformly distributed in $[0, 1]$. We know that $\mu = 1/2$. For $t \in [0, 1/2]$, we have

$$\mathbf{P}(|X - 1/2| > t) = \int_{x:|x-1/2|>t} f(x)dx = \int_0^{1/2-t} dx + \int_{1/2+t}^1 dx$$
$$= 1/2 - t + 1 - 1/2 - t = 1 - 2t.$$

We found that $\sigma = \sqrt{1/12} \approx 0.3$. Take $t = \sigma$. Hence

$$\mathbf{P}(|X - 1/2| > t) = 1 - 2t \sim 1 - 0.6 = 0.4.$$

This estimate outperforms Chebyshev's inequality.

The sharpness of the Chebyshev's inequality

Let us show that the Chebyshev inequality is a so-called "sharp estimate", i.e., it cannot be improved such that it is still true for all distributions of X. Let X be a random variable such that

$$X = \begin{cases} 1, & \text{with probability } 1/2 \\ -1, & \text{with probability } 1/2. \end{cases}$$

We have found that $\mu = 0$, $\operatorname{Var} X = 1$ and $\sigma = 1$. Consider $t \in [0, 1)$ only. Obviously,

$$\mathbf{P}(|X| > t) = 1.$$

On the other hand, the Chebyshev's inequality gives

$$\mathbf{P}(|X| > t) = \mathbf{P}(|X - \mu| > t) \leq \frac{1}{t^2} \to 1 \quad \text{as} \quad t \to 1.$$

The variance of the sum of independent random variables

Theorem 6.21 *Let X, Y be independent random variables. Then*

$$\operatorname{Var}(X + Y) = \operatorname{Var}(X) + \operatorname{Var}(Y).$$

Proof. Let $\overline{X} = X - \mathbf{E}X$ and $\overline{Y} = Y - \mathbf{E}Y$. It follows that $(X + Y) = \overline{X} + \overline{Y} + c$, where $c = \mathbf{E}X + \mathbf{E}Y$. Hence

$$\operatorname{Var}(X + Y) = \operatorname{Var}(\overline{X} + \overline{Y}).$$

Further,

$$\mathbf{E}\overline{X} = \mathbf{E}(X - \mathbf{E}X) = \mathbf{E}X - \mathbf{E}(\mathbf{E}X) = 0,$$
$$\mathbf{E}\overline{Y} = \mathbf{E}(X - \mathbf{E}Y) = \mathbf{E}Y - \mathbf{E}(\mathbf{E}Y) = 0.$$

Therefore

$$\operatorname{Var}(\overline{X} + \overline{Y}) = \mathbf{E}((\overline{Y} + \overline{X})^2) - (\mathbf{E}(\overline{X} + \overline{Y}))^2 = \mathbf{E}((\overline{Y} + \overline{X})^2)$$

and

$$\operatorname{Var}(X + Y) = \mathbf{E}((\overline{Y} + \overline{X})^2) = \mathbf{E}((\overline{Y})^2 + 2\mathbf{E}(\overline{X} \cdot \overline{Y}) + \mathbf{E}(\overline{X})^2).$$

Since $\overline{X}$ and $\overline{Y}$ are independent,

$$\mathbf{E}(\overline{X} \cdot \overline{Y}) = \mathbf{E}(\overline{X})(\mathbf{E}\overline{Y}) = 0,$$

and

$$\operatorname{Var} \overline{X} = \operatorname{Var} X, \quad \operatorname{Var} \overline{Y} = \operatorname{Var} Y.$$

Hence

$$\begin{aligned} \operatorname{Var}(X + Y) = \mathbf{E}((\overline{Y})^2 + 2\mathbf{E}(\overline{XY}) + \mathbf{E}(\overline{X})^2) &= \operatorname{Var} \overline{X} + \operatorname{Var} \overline{Y} \\ &= \operatorname{Var} X + \operatorname{Var} Y. \end{aligned}$$

Corollary 6.22 *Let $X_1, ..., X_n$ be independent random variables. Then*

$$\text{Var}\,(X_1 + ... + X_n) = \text{Var}\,(X_1) + ... + \text{Var}\,(X_n).$$

Proof. Let $S_k = X_1 + ... + X_k$. We have

$$\text{Var}\,(X_1 + ... + X_n) = \text{Var}\,(S_{n-1} + X_n) = \text{Var}\,S_{n-1} + \text{Var}\,X_n$$
$$= \text{Var}\,(S_{n-2} + X_{n-1}) + \text{Var}\,X_n = \text{Var}\,S_{n-2} + \text{Var}\,X_{n-1} + \text{Var}\,X_n = \dots$$

Repeating this, we obtain the desired statement.

6.5 Example: Binomial distribution

Let X belong to binomial distribution with parameters p and n. (We denote it as $Bin(p, n)$.) We have that $X = X_1 + ... + X_n$, where X_i are independent Bernoulli random variables

$$X_i = \begin{cases} 1, & \text{with probability } p \\ 0, & \text{with probability } 1 - p. \end{cases}$$

Previously, we found that $\text{Var}\,X_i = p(1-p)$. Hence

$$\text{Var}\,X = \text{Var}\,(X_1) + ... + \text{Var}\,(X_n) = n\text{Var}\,X_i = np(1-p).$$

Some analysis

We found that $\mathbf{E}X_i = p$ and $\mathbf{E}X = \mathbf{E}X_1 + ... + \mathbf{E}X_n = np$. Let $M_n = \frac{1}{n}X = \frac{1}{n}(X_1 + ... + X_n)$. Therefore

$$\mathbf{E}M_n = \frac{1}{n}\mathbf{E}X = p,$$

and

$$\text{Var}\,M_n = \frac{1}{n^2}\text{Var}\,X = \frac{np(1-p)}{n^2} = \frac{p(1-p)}{n} \to 0 \quad \text{as} \quad n \to +\infty.$$

Thus

$$\text{Var}\,M_n = \frac{1}{n^2}\text{Var}\,X \to 0 \quad \text{as} \quad n \to +\infty.$$

We obtained a very important result: the variance vanishes for $\overline{X} = X/n$ as $n \to +\infty$.

6.6 Covariance

Definition 6.23 *If X and Y are jointly distributed random variables with the expectations μ_X and μ_Y respectively, the covariance of X and Y is*

$$\mathrm{Cov}(X, Y) = \mathbf{E}((X - \mu_X)(Y - \mu_Y)).$$

The covariance of two random variables is a measure of their degree of association.

Some properties

(a) $\mathrm{Cov}(X, X) = \mathrm{Var}\, X$.
(b) $\mathrm{Cov}(X, Y) = \mathrm{Cov}(Y, X)$.
(c) $\mathrm{Cov}(-X, X) = -\mathrm{Var}\, X$. Note that, in this case, the covariance is negative, indicating a negative relationship between X and $-X$.
(d) $\mathrm{Cov}(X, Y) = \mathbf{E}(XY) - \mu_X \mu_y$.
(e) If Y and X are independent, then $\mathrm{Cov}(X, Y) = 0$.

Proof of (c): Let $Y = -X$. Then $\mu_Y = -\mu_X$ and

$$\begin{aligned}\mathrm{Cov}(-X, X) = \mathrm{Cov}(Y, X) &= \mathbf{E}((X - \mu_X)(Y - \mu_Y)) \\ &= \mathbf{E}((X - \mu_X)(-X + \mu_Y)) = -\mathrm{Var}\, X.\end{aligned}$$

Proof of (d):

$$\begin{aligned}\mathrm{Cov}(X, Y) &= \mathbf{E}((X - \mu_X)(Y - \mu_Y)) = \mathbf{E}(XY) - \mu_X \mathbf{E}Y - \mathbf{E}X \mu_Y + \mu_X \mu_Y \\ &= \mathbf{E}(XY) - \mu_X \mu_Y - \mu_X \mu_Y + \mu_X \mu_Y = \mathbf{E}(XY) - \mu_X \mu_Y.\end{aligned}$$

Proof of (e):

$$\mathrm{Cov}(X, Y) = \mathbf{E}(XY) - \mu_X \mu_Y = \mu_X \mu_Y - \mu_X \mu_Y = 0.$$

6.7 Covariance and linear transformations

Theorem 6.24 *Let X and Y be random variables, and $a \in \mathbf{R}$. Then*

$$\mathrm{Cov}(a + X, Y) = \mathrm{Cov}(X, Y).$$

Proof.

$$\begin{aligned}\mathrm{Cov}(a + X, Y) &= \mathbf{E}(a + X - \mathbf{E}(a + X))(Y - \mathbf{E}Y) \\ &= \mathbf{E}(a + X - a - \mathbf{E}X)(Y - \mathbf{E}Y) \\ &= \mathbf{E}(X - \mathbf{E}X)(Y - \mathbf{E}Y) \\ &= \mathrm{Cov}(X, Y).\end{aligned}$$

Example 6.25 Let $p \in [0, 1]$ be given and X be a Bernoulli random variable such that

$$X = \begin{cases} 1, & \text{with probability } p \\ 0, & \text{with probability } 1 - p. \end{cases}$$

Given $Y = 98 + X$, $\text{Cov}(X, Y) = \text{Cov}(X, X) = \text{Var}\, X = p(1 - p)$.

Theorem 6.26 *Let X, Y be random variables and $a, b \in \mathbf{R}$. Then*

$$\text{Cov}(aX, bY) = ab\text{Cov}(X, Y).$$

Proof.

$$\begin{aligned} \text{Cov}(aX, bY) &= \mathbf{E}(aX - \mathbf{E}(aX))(bY - \mathbf{E}(bY)) \\ &= ab\mathbf{E}(X - \mathbf{E}X)(Y - \mathbf{E}Y) = ab\text{Cov}(X, Y). \end{aligned}$$

Theorem 6.27 *Let X, Y, Z be random variables. Then*

$$\text{Cov}(X, Y + Z) = \text{Cov}(X, Y) + \text{Cov}(X, Z).$$

Proof.

$$\begin{aligned} \text{Cov}(X, Y + Z) &= \mathbf{E}(X - \mathbf{E}(X))(Y - \mathbf{E}(Y) + Z - \mathbf{E}(Z)) \\ &= \mathbf{E}((X - \mathbf{E}X)(Y - \mathbf{E}Y)) + \mathbf{E}((X - \mathbf{E}X)(Z - \mathbf{E}Z)) \\ &= \text{Cov}(X, Y) + \text{Cov}(X, Z). \end{aligned}$$

Corollary 6.28

$$\text{Var}\,(X + Y) = \text{Var}\, X + \text{Var}\, Y + 2\text{Cov}(X, Y).$$

Proof.

$$\begin{aligned} \text{Var}\,(X + Y) &= \text{Cov}(X + Y, X + Y) = \text{Cov}(X, X + Y) + \text{Cov}(Y, X + Y) \\ &= \text{Cov}(X, X) + \text{Cov}(X, Y) + \text{Cov}(Y, X) + \text{Cov}(Y, Y) \\ &= \text{Var}\, X + 2\text{Cov}(X, Y) + \text{Var}\, Y. \end{aligned}$$

6.8 Correlation

Definition 6.29 *If X and Y are jointly distributed random variables with standard deviations σ_X and σ_Y, respectively, the correlation coefficient of X and Y is*

$$\rho = \text{corr}(X, Y) = \frac{\text{Cov}(X, Y)}{\sigma_X \sigma_Y}.$$

Some properties of the correlation

(a) If Y and X are independent, then $\text{corr}(X, Y) = 0$.
(b) $\text{corr}(X, X) = 1$.
(c) $\text{corr}(X, Y) = \text{corr}(Y, X)$.
(d) $\text{corr}(-X, X) = -1$. In this case, the correlation is negative, indicating a negative relationship between X and $-X$.
(e) $\text{Var}\,(X + Y) = \text{Var}\,X + 2\text{corr}(X, Y)\sigma_X\sigma_Y + \text{Var}\,Y$.
(f) If $a, b \in \mathbf{R}$, $b \neq 0$, then $\text{corr}(a + bX, Y) = \frac{b}{|b|}\text{corr}(X, Y)$.

Proof of (f):

$$\begin{aligned}\text{corr}(a + bX, Y) &= \frac{\text{Cov}(a + bX, Y)}{\sqrt{\text{Var}\,(a + bX)}\sigma_Y} = \frac{b\text{Cov}(X, Y)}{\sqrt{b^2\text{Var}\,(X)} \cdot \sigma_Y} \\ &= \frac{b\text{Cov}(X, Y)}{|b|\sqrt{\text{Var}\,(X)} \cdot \sigma_Y} = \frac{b}{|b|}\text{corr}(X, Y).\end{aligned}$$

Theorem 6.30 *For any random variables* X, Y, $\text{corr}(X, Y) \in [-1, 1]$.

Proof. Let $\rho = \text{corr}(X, Y)$. First, consider the case where $\sigma_X = \sigma_Y = 1$. In this case, we have

$$\text{Var}\,(X + Y) = 1 + 2\rho + 1.$$

Since

$$\text{Var}\,(X + Y) \geq 0,$$

it follows that $\rho \geq -1$. Further, we have that

$$\text{Var}\,(X - Y) = 1 - 2\rho + 1.$$

Likewise, since

$$\text{Var}\,(X - Y) \geq 0,$$

it follows that $\rho \leq 1$.

For the general case, where $\sigma_X \neq 1$ or $\sigma_Y \neq 1$, observe that

$$\text{corr}(X, Y) = \text{corr}\left(\frac{X}{\sigma_X}, \frac{Y}{\sigma_Y}\right)$$

and

$$\text{Var}\left(\frac{X}{\sigma_X}\right) = \text{Var}\left(\frac{Y}{\sigma_Y}\right) = 1.$$

Apply the previous proof to the random variables $\overline{X} = \frac{X}{\sigma_X}$ and $\overline{Y} = \frac{Y}{\sigma_Y}$, Then the proof follows.

Corollary 6.31 *Assume that the marginal distributions of X, Y are given, but their correlation is variable. The minimum* $\operatorname{Var}(X+Y)$ *is achieved for the minimal value of* $\rho = \operatorname{corr}(X, Y)$.

For $\rho = -1$, we have

$$\begin{aligned}\operatorname{Var}(X+Y) &= \operatorname{Var} X - 2\sqrt{\operatorname{Var} X \operatorname{Var} Y} + \operatorname{Var} Y \\ &= \sigma_X^2 - 2\sigma_X\sigma_Y + \sigma_Y^2 = (\sigma_X - \sigma_Y)^2.\end{aligned}$$

6.9 Example: Investment portfolios

Return of an investment is

$$R = \frac{X}{X_0},$$

where X is the terminal wealth, X_0 is the original wealth. Assume that there are 3 companies that can ensure returns R_1, R_2, R_3 respectively. Due to uncertainty, the returns are random. An investor has to select 2 companies to invest \$1M in each of the two companies and it is known that

$$\operatorname{corr}(R_1, R_2) = 1/2, \quad \operatorname{corr}(R_2, R_3) = 0, \quad \operatorname{corr}(R_1, R_3) = -1/2.$$

Given that $\mathbf{E}R_k = a$ and $\operatorname{Var} R_k = \sigma^2$ for some $a > 0$, $\sigma >$, and for all k. The terminal wealth are

$$X = R_1 + R_2, \quad \text{or} \quad X = R_2 + R_3, \quad \text{or} \quad X = R_1 + R_3,$$

corresponding to the choice.

In practice, one wishes to maximize $\mathbf{E}X$ and minimize the risk, or minimize $\operatorname{Var} X$.

Obviously, $\mathbf{E}X = 2a$ for any choice.

However,

$$\begin{aligned}\operatorname{Var}(R_1+R_2) &= \sigma^2\left(2 + 2\cdot\frac{1}{2}\right) = 3\sigma^2, \\ \operatorname{Var}(R_2+R_3) &= \sigma^2\left(2 + 0\cdot\frac{1}{2}\right) = 2\sigma^2, \\ \operatorname{Var}(R_3+R_1) &= \sigma^2\left(2 - 2\cdot\frac{1}{2}\right) = \sigma^2.\end{aligned}$$

Therefore the last selection is optimal.

6.10 Example: Linearly related random variables

Let X, W be independent random variables such that $\mathbf{E}X = \mathbf{E}W = 0$. Let $a \in \mathbf{R}$, and let

$$Y = aX + W.$$

In many models, W is interpreted as a noise or an error. We have

$$\begin{aligned}\mathrm{Cov}(X,Y) &= \mathbf{E}(XY) = \mathbf{E}(aX^2 + XW)\\ &= a\mathbf{E}X^2 + \mathbf{E}(XW) = a\mathrm{Var}\,X + 0 = a\sigma_X^2\end{aligned}$$

and

$$\begin{aligned}\sigma_Y^2 = \mathrm{Var}\,Y &= \mathbf{E}((aX+W)^2) = \mathbf{E}(a^2X^2 + 2aXW + W^2)\\ &= a^2\mathbf{E}X^2 + 2a\mathbf{E}(XW) + \mathbf{E}W^2\\ &= a^2\sigma_X^2 + \sigma_W^2.\end{aligned}$$

Hence

$$\mathrm{corr}(X,Y) = \frac{\mathrm{Cov}(X,Y)}{\sigma_X\sqrt{a^2\sigma_X^2 + \sigma_W^2}} = \frac{a\sigma_X}{\sqrt{a^2\sigma_X^2 + \sigma_W^2}}.$$

In particular, $\mathrm{corr}(X,Y) = \pm 1$ only if $\sigma_W = 0$, i.e., if $\mathrm{Var}\,W = 0$, or if W is non-random.

Let us generalize it as follows.

Theorem 6.32 $\mathrm{corr}(X,Y) = \pm 1$ *if and only if there are non-random constants* $a, b \in \mathbf{R}$ *such that*

$$Y = aX + b.$$

Proof. Let us consider first the case where $\sigma_X = \sigma_Y = 1$. In this case,

$$\mathrm{Var}\,(X+Y) = 1 + 2\rho + 1.$$

Clearly, $\mathrm{Var}\,(X+Y) = 0$ if and only if $\rho = -1$. In addition, we have that $\mathrm{Var}\,(X+Y) = 0$ if and only if $\mathbf{P}(X+Y=b) = 1$, where $b \in \mathbf{R}$ ia a constant. Further,

$$\mathrm{Var}\,(X-Y) = 1 - 2\rho + 1.$$

Clearly, $\mathrm{Var}\,(X-Y) = 0$ if and only if $\rho = 1$. Again, we have that $\mathrm{Var}\,(X-Y) = 0$ if and only if $\mathbf{P}(X-Y=b) = 1$, where $b \in \mathbf{R}$ ia a constant.

For the general case, where $\sigma_X \neq 1$ or $\sigma_Y \neq 1$, observe again that

$$\mathrm{corr}(X,Y) = \mathrm{corr}\left(\frac{X}{\sigma_X}, \frac{Y}{\sigma_Y}\right),$$

and

$$\mathrm{Var}\left(\frac{X}{\sigma_X}\right) = \mathrm{Var}\left(\frac{Y}{\sigma_Y}\right) = 1.$$

Apply the previous proof to the random variables $\overline{X} = \frac{X}{\sigma_X}$ and $\overline{Y} = \frac{Y}{\sigma_Y}$. Then the proof follows.

Problems for Week 6

Problem 6.1 *Let X be a random variable with the normal distribution $N(0,4)$, and let Y be a random variable with the normal distribution $N(0,9)$. Assume that the correlation between X and Y is -0.5, i.e.* $\text{corr}(X,Y) = -0.5$. *Calculate* $\text{Var}(Z)$ *for* $Z = X + Y$.

Problem 6.2 *Let X be a random variable with the normal distribution $N(0,4)$, and let Y be a random variable with the normal distribution $N(0,9)$. Assume that the correlation between X and Y is -0.5, i.e.* $\text{corr}(X,Y) = -0.5$. *Calculate* $\text{Var}(Z)$ *for* $Z = X - Y$.

Problem 6.3 *Let X be a random variable with normal distribution $N(2,25)$, and let Y be a random variable with normal distribution $N(1,9)$. Assume that the correlation between X and Y is 0.5, i.e.* $\text{corr}(X,Y) = 0.5$. *Calculate* $\text{Var}(Z)$ *for* $Z = X - Y - 1$.

Problem 6.4 *Let Y be a random variable such that* $\mathbf{P}(Y = -1) = \mathbf{P}(Y = 1) = 1/2$ *and such that* $\text{corr}(X,Y) = 0.2$, *where X is the random variable such that* $\mathbf{E}X = 3$, $\text{Var}\,X = 2$. *Find* $\mathbf{E}Y$, $\text{Var}\,Y$, $\text{Cov}(X,Y)$ *and* $\mathbf{E}(XY)$.

Problem 6.5 *Let $X \sim U(-1,1)$, and let $Y = 1$ of $|X| < 1/2$, and $Y = 0$ if $|X| \geq 1/2$. Are these random variables independent? Find* $\text{Cov}(X,Y)$.

Problem 6.6 *Let X be a random variable with normal distribution $N(2,25)$ (i.e., with mean 2 and variance 25). Let Y be a random variable with normal distribution $N(1,9)$ (i.e., with mean 1 and variance 9). Assume that the correlation between X and Y is 0.5, i.e.* $\text{corr}(X,Y) = 0.5$. *Calculate* $\text{Var}(Z)$ *for* $Z = X - Y - 1$.

Problem 6.7 *Let $D = \{(x,y) : 0 \leq x \leq 1, 0 \leq y \leq x\}$. Let two random variables X and Y be such that the random vector (X,Y) is uniformly distributed in D. Find* $\mathbf{E}X$, $\mathbf{E}Y$, $\text{Cov}(X,Y)$.

Week 7. Conditional Expectations

In this chapter, we discuss conditional expectation of random variables and the applications for estimation of the parameters for random variables.

7.1 Conditional expectations given an observed value

Definition 7.1 *Let X and Y be discrete jointly distributed random variables with joint frequency $p(x,y)$. The conditional expectation of Y given $X = x$ is*

$$\mathbf{E}(Y|X=x) = \sum_y y p_{Y|X}(y|x).$$

Definition 7.2 *Let X and Y be jointly distributed random variables with joint density $f(x,y)$. The conditional expectation of Y given $X = x$ is*

$$\mathbf{E}(Y|X=x) = \int_{-\infty}^{\infty} y f_{Y|X}(y|x)dy.$$

Example 7.3 Consider again the example where a fair coin is tossed three times; X is the number of heads on the first toss and Y is the total number of heads. The joint frequency function $p(x,y)$ of X and Y is as given in the following table:

$x \backslash y$	0	1	2	3
0	1/8	1/4	1/8	0
1	0	1/8	1/4	1/8

We have that $p_X(0) = p_X(1) = 1/2$ and $p_{Y|X}(y|x) = p(x,y)/p(x)$. It gives the table for $p_{Y|X}(y|0) = p(0,y)/p(0)$:

$x\backslash y$	0	1	2	3
0	1/4	1/2	1/4	0

.

Hence

$$\mathbf{E}(Y|X=0) = 0 \cdot 1/4 + 1 \cdot 1/2 + 2 \cdot 1/4 + 3 \cdot 0 = 1.$$

Similarly, we obtain the table for $p_{Y|X}(y|1) = p(1,y)/p(0)$:

$x\backslash y$	0	1	2	3
1	0	1/4	1/2	1/4

Therefore

$$\mathbf{E}(Y|X=1) = 1 \cdot 1/4 + 2 \cdot 1/2 + 3 \cdot 1/4 + 3 \cdot 0 = 2.$$

Some properties

(a) If X, Y are independent then $\mathbf{E}(Y|X=x) = \mathbf{E}X$ for all x.
Proof. For continuous case, we have that $f_{Y|X}(y|x) = f_Y(x)$. Hence

$$\mathbf{E}(Y|X=x) = \int_{-\infty}^{\infty} y f_{Y|X}(y|x)dy = \int_{-\infty}^{\infty} y f_Y(y)dy.$$

(b) Let $g : \mathbf{R}^2 \to \mathbf{R}$ be a function and $Z = g(X, Y)$. Then

$$\mathbf{E}(Z|X=x) = \mathbf{E}(g(x,Y)|X=x).$$

For the continuous case, the last equality can be rewritten as:

$$\mathbf{E}(g(X,Y)|X=x) = \mathbf{E}(g(x,Y)|X=x) = \int_{-\infty}^{\infty} g(x,y) f_{Y|X}(y|x)dy.$$

Example 7.4 For $Z = X + Y$,

$$\mathbf{E}(Z|X=x) = \mathbf{E}(X+Y|X=x) = x + \mathbf{E}(Y|X=x).$$

For $Z = XY$,

$$\mathbf{E}(Z|X=x) = \mathbf{E}(XY|X=x) = x\mathbf{E}(Y|X=x).$$

For $Z = X^2 + Y^2$,

$$\mathbf{E}(Z|X=x) = \mathbf{E}(X^2+Y^2|X=x) = x^2 + \mathbf{E}(Y^2|X=x).$$

7.2 Conditional expectation as an expectation

In fact, conditional expectation is the expectation on a new probability space, for an observer for whom the probabilistic hypothesis is affected by his/her knowledge of the fact that $X = x$.

Therefore, the properties of established for expectations holds for conditional expectations as well.

For instance, the following properties can be mentioned:

(a) $\mathbf{E}(Y + Z|X = x) = \mathbf{E}(Y|X = x) + \mathbf{E}(Z|X = x)$, for three random variables.
(b) $\mathbf{E}(aY|X = x) = a\mathbf{E}(Y|X = x)$ for $a \in \mathbf{R}$.
(c) Finite $\mathbf{E}(Y|X = x)$ exists if

$$\sum_y |y| p_{Y|X}(y|x) < +\infty$$

(discrete case) or

$$\int_{-\infty}^{\infty} |y| f_{Y|X}(y|x) dy < +\infty$$

(continuous case).

All basic definitions (medians, moments, etc) can be extended on the case of conditional probability space.

7.3 Conditional moments

Let X, Y be random variables.

Definition 7.5 *The kth conditional moment of the random variable is defined as*

$$\mathbf{E}(X^k|Y = y).$$

By the rule of expectation, we have

$$\mathbf{E}(X^k|Y = y)) = \begin{cases} \sum_x x^k p_{Y|X}(x|y) & \text{if } X \text{ is discrete,} \\ \int_{-\infty}^{\infty} x^k f_{X|Y}(x|y) dx & \text{if } X \text{ is continuous.} \end{cases}$$

Definition 7.6 *The kth conditional central moment (or the moment about the mean) of the random variable X is defined as*

$$\mathbf{E}((X - \mu(y))^k|Y = y)],$$

where $\mu(y) = \mathbf{E}(X|Y = y)$.

By the rule of expectation, we have

$$\begin{cases} \sum_x (x-\mu(y))^k p_{Y|X}(x|y) & \text{if } X \text{ is discrete} \\ \int_{-\infty}^{\infty} (x-\mu(y))^k p_{Y|X}(x|y) dx & \text{if } X \text{ is continuous.} \end{cases}$$

Conditional variance

Definition 7.7 *The conditional variance given $Y = y$ is*

$$\mathbf{E}((X-\mu(y))^2|Y=y)),$$

where $\mu(y) = \mathbf{E}(X|Y=y)$.

In fact, it is the conditional 2nd central moment defined above.

Theorem 7.8

$$\mathbf{E}((X-\mu(y))^2|Y=y)) = \mathbf{E}((X)^2|Y=y)) - (\mathbf{E}(X|Y=y))^2.$$

Example 7.9 Consider again the example where a fair coin is tossed three times; X is the number of heads on the first toss, and Y is the total number of heads. We have the table for $p_{Y|X}(y|0) = p(0,y)/p(0)$

$x \backslash y$	0	1	2	3
0	1/4	1/2	1/4	0

We calculated

$$\mathbf{E}(Y|X=0) = 1.$$

This gives us

$$\mathbf{E}(Y^2|X=0) = 0 \cdot 1/4 + 1^2 \cdot 1/2 + 2^2 \cdot 1/4 + 3 \cdot 0 = 3/2.$$

Hence, the conditional variance given $X = 0$ is $3/2 - 1^2 = 1/2$.

7.4 Conditional expectation as a random variable

Before, we considered expectations as non-random characteristics of random variables; $\mathbf{E}(X)$ was a non-random value assigned to random variable X. Below, we consider another type of conditional expectations that are random variables.

Definition 7.10 *Let X and Y be jointly distributed random variables. The conditional expectation of Y given $X = x$ is actually a value defined for any x. Hence, it can be considered as a function $h(x) = \mathbf{E}(Y|X=x)$. The random variable $h(X)$ is said to be the conditional expectation of Y given X. It is denoted as $\mathbf{E}(Y|X)$.*

Example 7.11 Consider again the example where a fair coin is tossed three times; X is the number of heads on the first toss and Y is the total number of heads. We found that

$$\mathbf{E}(Y|X=0)=1, \quad \mathbf{E}(Y|X=1)=2.$$

Therefore, $\mathbf{E}(Y|X)=h(X)$, where

$$h(x)=\begin{cases}1, & x=0\\ 2, & x=1.\end{cases}$$

It can be also seen that $\mathbf{E}(Y|X)$ is a random variable such that

$$\mathbf{E}(Y|X)=\begin{cases}1, & \text{with probability } 1/2\\ 2, & \text{with probability } 1/2.\end{cases}$$

Random conditional expectation as an expectation

Again, a random variable represented as the conditional expectation $\mathbf{E}(Y|X)$ can be interpreted as the expectation on a new probability space, for an observer for whom the probabilistic hypothesis is affected by his/her knowledge of the value of X.

Therefore, the properties of established for expectations holds for conditional expectations as well.

The random event that $\mathbf{E}(Y|X)$ exists and is finite occurs if

$$\sum_y |y| p_{Y|X}(y|X) < +\infty$$

(discrete case) or

$$\int_{-\infty}^{\infty} |y| f_{Y|X}(y|X)dy < +\infty$$

(continuous case).

All basic definitions (medians, moments, etc) can be extended on the case of conditional probability space. The corresponding characteristics will be random variables that can be represented as functions of X.

The following properties can be mentioned:

(a) $\mathbf{E}(Y+Z|X)) = \mathbf{E}(Y|X)+\mathbf{E}(Z|X)$ for three random variables.
(b) $\mathbf{E}(aY|X)=a\mathbf{E}(Y|X)$ for $a\in\mathbf{R}$.
(c) If X, Y are independent then $\mathbf{E}(X|Y)=\mathbf{E}X$.

Theorem 7.12

$$\mathbf{E}(\mathbf{E}(X|Y))=\mathbf{E}X.$$

Proof.

$$\mathbf{E}(\mathbf{E}(X|Y)) = \mathbf{E}h(Y),$$

where

$$h(y) = \mathbf{E}(X|Y = y).$$

We consider the case of continuous distributions only. Let $f(x, y)$ be the joint density for (X, Y). We have

$$\mathbf{E}(\mathbf{E}(X|Y)) = \mathbf{E}h(Y) = \int_{-\infty}^{\infty} h(y) f_Y(y)dy = \int_{-\infty}^{\infty} \mathbf{E}(X|Y = y) f_Y(y)dy$$

$$= \int_{-\infty}^{\infty} dx \left(\int_{-\infty}^{\infty} x f_{X|Y}(x|y)dx \right) f_Y(y)dy$$

$$= \int_{-\infty}^{\infty} x dx \int_{-\infty}^{\infty} f_{X|Y}(x|y) f_Y(y)dy = \int_{-\infty}^{\infty} x dx \int_{-\infty}^{\infty} f(x, y)dy = \mathbf{E}X.$$

Example 7.13 Consider again the example where a fair coin is tossed three times; X is the number of heads on the first toss, and Y is the total number of heads. We found that $\mathbf{E}(Y|X)$ is a random variable such that

$$\mathbf{E}(Y|X) = \begin{cases} 1, & \text{with probability } 1/2 \\ 2, & \text{with probability } 1/2. \end{cases}$$

Hence $\mathbf{E}(\mathbf{E}(Y|X)) = 1 \cdot 1/2 + 2 \cdot 1/2 = 3/2$. It can also be calculated directly that $\mathbf{E}Y = 3/2$.

An important example: Random sums

Let N be a random variable with integer nonnegative values, and let X_k be random variables. We assume that N is independent from all X_k and that $\mathbf{E}N = \nu$ and $\mathbf{E}X_k = \mu$ for all k. Let

$$Y = \sum_{k=1}^{N} X_k.$$

Let us find $\mathbf{E}(Y)$. We have

$$\mathbf{E}(Y) = \mathbf{E} \sum_{k=1}^{N} X_k = \mathbf{E}\left(\mathbf{E}\left(\sum_{k=1}^{N} X_k | N \right)\right).$$

Here $\mathbf{E}\left(\sum_{k=1}^{N} X_k | N \right)$ is the expectation for an observer who knows the value of N; it is non-random for him/her. In addition, $\mathbf{E}X_k = \mathbf{E}(X_k|N) = \mu$. Hence,

$$\mathbf{E}\left(\sum_{k=1}^{N} X_k | N \right) = \sum_{k=1}^{N} \mathbf{E}(X_k|N) = N\mathbf{E}X_k = N\mu.$$

It gives

$$\mathbf{E}Y = \mathbf{E}(N\mu) = (\mathbf{E}N)\mu = \nu\mu.$$

7.5 Expectation as the best estimate

Theorem 7.14 *Let X be such that $\mathbf{E}(X^2) < +\infty$ and $\mu = \mathbf{E}(X)$. Then*

$$\mathbf{E}(|X-\mu|^2) \le \mathbf{E}(|X-c|^2) \quad \textit{for all} \quad c \in \mathbf{R}.$$

Proof. Clearly,

$$\begin{aligned}\mathbf{E}((X-c)^2) &= \mathbf{E}(X-\mu+\mu-c)^2 \\ &= \mathbf{E}(X-\mu)^2 + 2c\mathbf{E}(X-\mu) + (\mathbf{E}X-c)^2 \\ &= \mathbf{E}(X-\mu)^2 + (\mu-c)^2 \ge \mathbf{E}(X-\mu)^2.\end{aligned}$$

In this sense, the expectation $\mathbf{E}(X)$ is the best non-random estimate (prediction) of X.

Theorem 7.15 *Let X, Y be jointly distributed random variables, $\mathbf{E}(X^2) < +\infty$. Then*

$$\mathbf{E}(|X - \mathbf{E}(X|Y)|^2) \le \mathbf{E}(|X - h(Y)|^2)$$

for all non-random functions $h : \mathbf{R} \to \mathbf{R}$.

For this theorem, the conditional expectation $\mathbf{E}(X|Y)$ is the best estimate of X based on the observations of Y.

Example 7.16 Assume that $X = aY + \varepsilon$, where $a \in \mathbf{R}$ is known; X, Y, ε are random variables; Y and ε are independent with $\mathbf{E}\varepsilon = 0$. Calculate $\mathbf{E}(X|Y)$.

In this case, ε can be interpreted as an error; Y is an observation, X is a variable to be estimated by the observation of Y.

Hence

$$\mathbf{E}(X|Y) = \mathbf{E}(aY + \varepsilon|Y) = \mathbf{E}(aY|Y) + \mathbf{E}(\varepsilon|Y) = aY + \mathbf{E}\varepsilon = aY.$$

We use here that $\mathbf{E}(\varepsilon|Y) = \mathbf{E}\varepsilon$, by independence of Y and ε. It is the best possible estimate of X given that Y is known.

Example 7.17 Assume that $X = aY + b + \varepsilon$, where $a, b \in \mathbf{R}$ are known; X, Y, ε are random variables; Y and ε are independent with $\mathbf{E}\varepsilon = 0$. Calculate $\mathbf{E}(X|Y)$.

We have

$$\begin{aligned}\mathbf{E}(X|Y) = \mathbf{E}(aY + b + \varepsilon|Y) &= \mathbf{E}(aY|Y) + b + \mathbf{E}(\varepsilon|Y)\\ &= aY + b + \mathbf{E}\varepsilon = aY + b.\end{aligned}$$

We use here that $\mathbf{E}(\varepsilon|Y) = \mathbf{E}\varepsilon$, by independence of Y and ε.

Again, it is the best possible estimate of X given that Y is known.

Example 7.18 Consider the sample space Ω has three elements only, i.e., $\Omega = \{\omega_1, \omega_2, \omega_3\}$. We assume that the set of random events is 2^Ω and that

$$\mathbf{P}(\{\omega_1\}) = \tfrac{1}{4}, \quad \mathbf{P}(\{\omega_2\}) = \tfrac{1}{4}, \quad \mathbf{P}(\{\omega_3\}) = \tfrac{1}{2}.$$

Assume that $\xi(\omega_i) = i - 2$ and $\eta = |\xi|$. Let us find $\mathbf{E}\{\xi\,|\,\eta\}$ and express $\mathbf{E}\{\xi\,|\,\eta\}$ as a deterministic function of η.

We have

$$\begin{aligned}\eta(\omega) &= 0, \quad \omega = \omega_2,\\ \eta(\omega) &= 1, \quad \omega = \omega_1 \quad \text{or} \quad \omega = \omega_3.\end{aligned}$$

It follows that $\mathbf{E}\{\xi\,|\,\eta\} = \hat{\zeta}$, where $\hat{\zeta} = \hat{h}(\eta)$ for some function $\hat{h} : \mathbf{R} \to \mathbf{R}$ such that $\mathbf{E}(\zeta - \xi)^2$ is minimal over functions $h : \mathbf{R} \to \mathbf{R}$. Any function $\zeta = h(\eta)$ has a form $\zeta(\omega) = \alpha$ for $\omega = \omega_2$ and $\zeta(\omega) = \beta$ for $\omega = \omega_1$ or $\omega = \omega_3$, $\alpha, \beta \in \mathbf{R}$.

Let $f(\alpha, \beta) = \mathbf{E}(\zeta - \xi)^2$. It suffices to find (α, β) such that $f(\alpha, \beta) = \mathbf{E}(\zeta - \xi)^2 = \min$. Hence

$$\begin{aligned}f(\alpha, \beta) &= \mathbf{E}(\zeta - \xi)^2\\ &= \mathbf{P}(\{\omega_1\})(\beta + 1)^2 + \mathbf{P}(\{\omega_2\})(\alpha - 0)^2 + \mathbf{P}(\{\omega_3\})(\beta - 1)^2\\ &= \frac{1}{4}(\beta + 1)^2 + \frac{1}{4}\alpha^2 + \frac{1}{2}(\beta - 1)^2,\end{aligned}$$

and

$$\frac{\partial f}{\partial \alpha}(\alpha, \beta) = \frac{\alpha}{2}, \quad \frac{\partial f}{\partial \beta}(\alpha, \beta) = \frac{1}{2}(\beta + 1) + \beta - 1 = \frac{3}{2}\beta - \frac{1}{2}.$$

Therefore, the only minimum of f is at $(\alpha, \beta) = (0, 1/3)$ and

$$\begin{aligned}\mathbf{E}\{\xi\,|\,\eta\}(\omega) &= 0, \quad \omega = \omega_2,\\ \mathbf{E}\{\xi\,|\,\eta\}(\omega) &= 1/3, \quad \omega = \omega_1 \quad \text{or} \quad \omega = \omega_3.\end{aligned}$$

Clearly, $\mathbf{E}\{\xi\,|\,\eta\}$ can be expressed as a deterministic function of η:

$$\mathbf{E}\{\xi\,|\,\eta\} = h(\eta), \quad \text{where} \quad h(0) = 0, \quad h(x) = 1/3, \quad x \neq 0.$$

Problems for Week 7

Problem 7.1 *Assume that* $X = a + bY + cY^2 + \varepsilon$, *where* $a, b, c \in \mathbf{R}$ *are known,* X, Y, ε *are random variables such that* Y *and* ε *are independent, and that* $\mathbf{E}\varepsilon = 0$. *Calculate* $\mathbf{E}(X|Y)$.

Problem 7.2 *Assume that* $X^{1/3} = Y + \varepsilon$, *where* X, Y, ε *are random variables such that* Y *and* ε *are independent, and that* $\varepsilon \sim U(0,1)$. *Calculate* $\mathbf{E}(X|Y)$.

Problem 7.3 *Consider the sample space* Ω *has three elements only, i.e.,* $\Omega = \{\omega_1, \omega_2, \omega_3\}$. *We assume that the set of random events is* 2^Ω *and that*

$$\mathbf{P}(\{\omega_1\}) = 1/4, \quad \mathbf{P}(\{\omega_2\}) = 1/4, \quad \mathbf{P}(\{\omega_3\}) = 1/2.$$

Let $\xi(\omega_1) = 0.1$, $\xi(\omega_2) = 0$, $\xi(\omega_3) = -0.1$. *Let* $\eta(\omega_1) = \eta(\omega_3) = 0.1$, $\eta(\omega_2) = -0.1$. *Find* $\mathbf{E}\{\xi|\eta\}$ *and express* $\mathbf{E}\{\xi \mid \eta\}$ *as a deterministic function of* η.

Week 8. Moment Generating Functions

In this chapter, we discuss the moment generating functions and its applications to analysis of the probability distributions.

8.1 Definition of moment generating function

Definition 8.1 *Moment generating function (m.g.f.) of a random variable X is*

$$M(t) = \mathbf{E}e^{tX},$$

if this expectation is defined.

For discrete case,

$$M(t) = \sum_x e^{tx} p(x),$$

where $p(x)$ is the frequency.

For continuous case,

$$M(t) = \int_{-\infty}^{\infty} e^{tx} f(x) dx,$$

where $f(x)$ is the density.

Some properties: It follows from the definitions that $M(0) = 1$ and $M(t) \geq 0$ for all t where the m.g.f. is defined.

Example 8.2 Let X be a Bernoulli random variable such that

$$X = \begin{cases} 1, & \text{with probability } p \\ 0, & \text{with probability } 1 - p. \end{cases}$$

We have that $M(t) = \mathbf{E}e^{tX} = e^t \cdot p + 1 \cdot (1 - p) = pe^t + 1 - p$. This m.g.f. is defined for all t.

Example 8.3 Let X be uniformly distributed in $[0,1]$. Its m.g.f. is

$$M(t) = \mathbf{E}e^{tX} = \int_0^1 e^{tx}dx = \frac{1}{t}e^{tx}\Big|_0^1 = \frac{e^t - 1}{t}.$$

This m.g.f. is defined for all t.

Example 8.4 Let X be uniformly distributed in $[a,b]$. Then the m.g.f. is

$$M(t) = \mathbf{E}e^{tX} = \frac{1}{b-a}\int_a^b e^{tx}dx = \frac{1}{(b-a)t}e^{tx}\Big|_a^b = \frac{e^{bt} - e^{at}}{t(b-a)}.$$

This m.g.f. is defined for all t.

Example 8.5 Let X be distributed exponentially with parameter $\lambda > 0$, with the density

$$f(x) = \begin{cases} \lambda e^{-\lambda x}, & x \geq 0 \\ 0, & \text{otherwise.} \end{cases}$$

If $t < \lambda$, $t - \lambda < 0$ and the m.g.f. is

$$M(t) = \mathbf{E}e^{tX} = \int_0^\infty e^{tx}\lambda e^{-\lambda x}dx = \frac{\lambda}{t-\lambda}e^{(t-\lambda)x}\Big|_0^\infty = \frac{\lambda}{\lambda - t}.$$

If $t \geq \lambda$, the m.g.f. is not defined.

8.2 m.g.f. for a sum of independent random variables

Theorem 8.6 *Let X and Y be independent random variables with the m.g.f. $M_X(t)$ and $M_Y(t)$ defined for some t. $X+Y$ has the m.g.f. $M_{X+Y}(t)$ defined for the same t, where*

$$M_{X+Y}(t) = M_X(t)M_Y(t).$$

Proof. We have that

$$M_{X+Y}(t) = \mathbf{E}e^{t(X+Y)} = \mathbf{E}e^{tX+tY} = \mathbf{E}\left(e^{tX}e^{tY}\right) = e^{tX}\mathbf{E}e^{tY} = M_X(t)M_Y(t).$$

Example 8.7 Let X be the total number of successes in n Bernoulli trials. We have $X = X_1 + ... + X_n$, where X_i are Bernoulli variables such that

$$X_i = \begin{cases} 1, & \text{with probability } p \\ 0, & \text{with probability } 1-p. \end{cases}$$

From the independence of $\{X_i\}$, it follows that

$$M(t) = \mathbf{E}e^{tX} = \mathbf{E}\exp(t(X_1 + ... + X_n))$$
$$= \mathbf{E}\Big(\exp(tX_1) \times \cdots \times \exp(tX_n)\Big)$$
$$= \mathbf{E}\exp(tX_1) \times \cdots \times \mathbf{E}\exp(tX_n).$$

However,

$$\mathbf{E}e^{tX_i} = pe^t + 1 - p.$$

Therefore,

$$M(t) = (pe^t + 1 - p)^n.$$

Existence of m.g.f.

Theorem 8.8 *Let X be a discrete random variable. The m.g.f. $M(t)$ is defined for t such that*

$$\sum_x e^{tx}p(x)dx < +\infty,$$

where $p(x)$ is the frequency.

Theorem 8.9 *Let X be a continuous random variable. The m.g.f. $M(t)$ is defined for t such that*

$$\int_{-\infty}^{\infty} e^{tx} f(x)dx < +\infty,$$

where $f(x)$ is the density.

Corollary 8.10 *If $\mathbf{P}(|X| \leq M) = 1$ for some $M > 0$, then the m.g.f. $M(t)$ is defined for all t.*

Proof. Let us consider the discrete case. It suffices to show that $\forall t$, for the frequency $p(x)$,

$$\sum_x e^{tx}p(x) < +\infty.$$

If $|x| > M$ then $p(x) = 0$. Thus,

$$\sum_x e^{tx}p(x) \leq \sum_x e^{tM}p(x) \leq e^{tM}\sum_x p(x) \leq e^{tM} < +\infty.$$

For the continuous case, it suffices to show that $\forall t$, for the density f,

$$\int_{-\infty}^{\infty} e^{tx} f(x)dx < +\infty.$$

If $|x| > M$ then $f(x) = 0$. Thus,

$$\int_{-\infty}^{\infty} e^{tx} f(x)dx \leq \int_{-M}^{M} e^{tM} f(x)dx \leq e^{tM}\int_{-M}^{M} f(x)dx \leq e^{tM} < +\infty.$$

8.3 m.g.f. and linear transformation of the random variable

Theorem 8.11 *Let X be a random variable with the m.g.f. $M_X(t)$ and*

$$Y = a + bX,$$

where $a, b \in \mathbf{R}$. The m.g.f. for Y is

$$M_Y(t) = e^{at} M_X(bt).$$

Proof.

$$\begin{aligned} M_Y(t) &= \mathbf{E}e^{(a+bX)t} = \mathbf{E}e^{at+bXt} \\ &= \mathbf{E}e^{at}e^{bXt} = e^{at}\mathbf{E}e^{(bt)X}. \end{aligned}$$

Hence,

$$M_Y(t) = e^{at} M_X(bt).$$

Example 8.12 Let X be uniformly distributed in $[0, 1]$ and

$$Y = a + (b - a)X,$$

where $b > a$. The densities for X and Y are related as

$$f_Y(y) = \frac{1}{b-a} f_X\left(\frac{y-a}{b-a}\right).$$

In addition, we have

$$f_X(x) = \mathbb{I}_{[0,1]}(x).$$

Therefore,

$$f_Y(y) = \frac{1}{b-a}\mathbb{I}_{[a,b]}(y).$$

Thus, Y is uniformly distributed in $[a, b]$. We found that the m.g.f. for X is

$$M_X(t) = \frac{e^t - 1}{t}$$

and the m.g.f. for Y is

$$M_Y(t) = \frac{e^{bt} - e^{at}}{t(b-a)}.$$

This m.g.f. is defined for all t.

On the other hand, Theorem 8.11 gives that

$$M_X((b-a)t) = \frac{e^{(b-a)t} - 1}{t(b-a)}.$$

It follows that

$$M_Y(t) = e^{at} M_X((b-a)t).$$

Example 8.13 Let X be distributed exponentially with parameter $\lambda > 0$, with the density

$$f_X(x) = \mathbb{I}_{\{[0,\infty]\}}\lambda e^{-\lambda x} = \begin{cases} \lambda e^{-\lambda x}, & x \geq 0 \\ 0, & \text{otherwise.} \end{cases}$$

Consider $Y = a + X$. By Theorem 8.11, the m.g.f. for Y is

$$M_Y(t) = e^{at} M_X(t).$$

It can also be calculated directly using

$$f_Y(y) = f_X(y - a) = \mathbb{I}_{\{[a,\infty]\}}\lambda e^{-\lambda(x-a)} = \begin{cases} \lambda e^{-\lambda(x-a)}, & x \geq a \\ 0, & \text{otherwise.} \end{cases}$$

8.4 m.g.f. and the moments

Let us discuss first some properties of the exponent.

Lemma 8.14 *The Taylor series for e^{tx} is*

$$e^{tx} = 1 + tx + \frac{t^2x^2}{2!} + ... + \frac{t^kx^k}{k!} + ...$$

Theorem 8.15 *Let $X \geq 0$ be a random variable with the m.g.f. $M(t)$ such that $M(t)$ is defined for some $t > 0$. Then all moments $\mu'_k = \mathbf{E}(X^k)$ are defined and finite.*

Proof. We have

$$e^{tx} = 1 + tx + \frac{t^2x^2}{2!} + ... + \frac{t^kx^k}{k!} + ...$$

It follows that, for any $k \geq 0$,

$$\mathbf{E}e^{tX} = \mathbf{E}\left(1 + tX + \frac{t^2X^2}{2!} + ... + \frac{t^kX^k}{k!} + ...\right) < +\infty.$$

Therefore,

$$\mathbf{E}\frac{t^kX^k}{k!} = \frac{t^k}{k!}\mathbf{E}(X^k) < +\infty$$

Theorem 8.16 *Let $X \leq 0$ be a random variable with the m.g.f. $M(t)$ such that $M(t)$ is defined for some $t < 0$. Then all moments $\mu'_k = \mathbf{E}(X^k)$ are defined and finite.*

Proof. Let $Y = -X$. We have that $Y \geq 0$ and

$$M_Y(-t) = \mathbf{E}e^{-tY} = \mathbf{E}e^{tX} = M_X(t).$$

By the previous theorem, it follows that all moments $\mathbf{E}(X^k)$ are defined and finite. Hence, all moments

$$\mathbf{E}(Y^k) = (-1)^k \mathbf{E}(X^k)$$

are defined and finite.

Corollary 8.17 *The methods based on m.g.f. are applicable for random variables such that all their moments are defined (i.e., for the distributions without heavy tails).*

Lemma 8.18

$$\left.\frac{d^k e^{tx}}{dt^k}\right|_{t=0} = x^k.$$

Proof.

$$\left.\frac{d}{dt}e^{tx}\right|_{t=0} = \left. xe^{xt}\right|_{t=0} = x = x^1.$$

Hence the Lemma is proved for $k = 1$. Further,

$$\left.\frac{d^2}{dt^2}e^{tx}\right|_{t=0} = \left.\frac{d}{dt}(xe^{tx})\right|_{t=0} = \left. x^2 e^{tx}\right|_{t=0} = x^2.$$

It can be continued for $k = 3, 4, ...$

Theorem 8.19 *Let X be a random variable with the m.g.f. $M(t)$. Assume $\exists \varepsilon : \varepsilon > 0$ and $M(t)$ is defined $\forall t \in (-\varepsilon, \varepsilon)$. The moments $\mu_k' = \mathbf{E}(X^k)$ can be calculated as*

$$\mu_k' = \frac{d^k M}{dt^k}(0), \quad k = 1, 2, ...$$

Proof. Let us consider the case where X is continuous with p.d.f. $f(x)$ and

$$M(t) = \mathbf{E}e^{tX} = \int_{-\infty}^{\infty} e^{tx} f(x)dx.$$

We have

$$\frac{d^k M}{dt^k}(0) = \left.\left(\frac{d^k}{dt^k}\int_{-\infty}^{\infty} e^{tx} f(x)dx\right)\right|_{t=0}$$
$$= \left.\left(\int_{-\infty}^{\infty} \frac{d^k}{dt^k}e^{tx} f(x)dx\right)\right|_{t=0} = \int_{-\infty}^{\infty} \left.\left(\frac{d^k}{dt^k}e^{tx}\right)\right|_{t=0} f(x)dx.$$

By Lemma 8.18, we obtain

$$\frac{d^k M}{dt^k}(0) = \int_{-\infty}^{\infty} x^k f(x)dx = \mathbf{E}(X^k) = \mu_k',$$

which completes the proof.

8.5 Calculation of the moments using m.g.f.

Example 8.20 Let X be a Bernoulli random variable such that

$$X = \begin{cases} 1, & \text{with probability } p \\ 0, & \text{with probability } 1-p. \end{cases}$$

We found that $M(t) = \mathbf{E}e^{tX} = e^t \cdot p + 1 \cdot (1-p) = pe^t + 1 - p$. This m.g.f. is defined for all t. Therefore,

$$\mathbf{E}(X) = \frac{dM}{dt}(0) = pe^t \Big|_{t=0} = p,$$

$$\mathbf{E}(X^2) = \frac{d^2M}{dt^2}(0) = pe^t \Big|_{t=0} = p,$$

and

$$\mathbf{E}(X^k) = \frac{d^kM}{dt^k}(0) = pe^t \Big|_{t=0} = p, \quad k = 1, 2, 3, ...$$

Example 8.21 Let X be the total number of successes in $n > 2$ Bernoulli trials. We have that

$$M(t) = (pe^t + 1 - p)^n.$$

It gives

$$\mathbf{E}(X) = \frac{dM}{dt}(0) = n(pe^t + 1 - p)^{n-1}pe^t \Big|_{t=0} = np,$$

and

$$\begin{aligned} \mathbf{E}(X^2) &= \frac{d^2M}{dt^2}(0) = \frac{d}{dt}\left(n(pe^t + 1 - p)^{n-1}pe^t\right)\Big|_{t=0} \\ &= \left((n(n-1)(pe^t + 1 - p)^{n-2}p^2e^{2t} + n(pe^t + 1 - p)^{n-1}pe^t\right)\Big|_{t=0} \\ &= n(n-1)p^2 + np = np(np - p + 1) \\ &= np(1-p) + n^2p^2. \end{aligned}$$

Example 8.22 Let X be uniformly distributed in $[0, 1]$. The m.g.f. is

$$M(t) = \frac{e^t - 1}{t} = \frac{1}{t}\left((1 + t + \frac{t^2}{2!} + \frac{t^3}{3!} + ...) - 1\right) = \frac{1}{t}\left(t + \frac{t^2}{2!} + \frac{t^3}{3!} + ...\right).$$

It gives

$$M(t) = 1 + \frac{t}{2!} + \frac{t^2}{3!} + \frac{t^3}{4!} + \frac{t^4}{5!} + ...$$

This m.g.f. is defined for all t.

Theorem 8.19 gives

$$\mathbf{E}(X) = \frac{dM}{dt}(0) = \frac{1}{2},$$
$$\mathbf{E}(X^2) = \frac{d^2M}{dt^2}(0) = \frac{d^2}{dt^2}\frac{t^2}{3!}\bigg|_{t=0} = \frac{d}{dt}\frac{2t}{3!}\bigg|_{t=0} = \frac{2}{3!}\bigg|_{t=0} = \frac{1}{3},$$
$$\mathbf{E}(X^3) = \frac{d^3M}{dt^3}(0) = \frac{d^3}{dt^3}\frac{t^3}{4!}\bigg|_{t=0} = \frac{d^2}{dt^2}\frac{3t^2}{4!}\bigg|_{t=0} = \frac{d}{dt}\frac{6t}{4!}\bigg|_{t=0} = \frac{1}{4}.$$

Example 8.23 Let X be distributed exponentially with parameter $\lambda > 0$ and the density

$$f(x) = \begin{cases} \lambda e^{-\lambda x}, & x \geq 0 \\ 0, & \text{otherwise.} \end{cases}$$

For $t < \lambda$, the m.g.f. is

$$M(t) = \frac{\lambda}{\lambda - t}.$$

For $t \geq \lambda$, the m.g.f. is not defined. Theorem 8.19 gives

$$\mathbf{E}X = M'(0) = \frac{\lambda}{(\lambda - t)^2}\bigg|_{t=0} = \frac{1}{\lambda}.$$

Example 8.24 Further, we obtain

$$\mathbf{E}(X^2) = \frac{d^2M}{dt^2}(0) = \frac{d^2}{dt^2}\frac{\lambda}{\lambda - t}\bigg|_{t=0} = \frac{d}{dt}\frac{\lambda}{(\lambda - t)^2}\bigg|_{t=0} = \frac{2\lambda}{(\lambda - t)^3}\bigg|_{t=0} = \frac{2}{\lambda^2}.$$

Similarly, we obtain

$$\mathbf{E}(X^3) = \frac{d^3M}{dt^3}(0) = \frac{d^2}{dt^2}\frac{\lambda}{\lambda - t}\bigg|_{t=0} = \frac{d}{dt}\frac{2\lambda}{(\lambda - t)^3}\bigg|_{t=0} = \frac{6\lambda}{(\lambda - t)^4}\bigg|_{t=0} = \frac{6}{\lambda^3}.$$

8.6 On the polynomial and exponential m.g.f.

Theorem 8.25 *Let X be a random variable with the m.g.f. $M_X(t)$ such that*

$$M_X(t) = a_0 + a_1 t + a_2 t^2 + ... + a_n t^n,$$

for all $t \in (-\varepsilon, \varepsilon)$, where $\varepsilon > 0$, $n \geq 0$, $a_i \in \mathbf{R}$. Then $X = 0$. In addition, $a_0 = 1$ and $a_i = 0$ for $i > 0$.

Proof. Let $m > n$ be an even number (for instance, take $m = 2n + 2$). We have that

$$\frac{d^m M}{dt^m}(0) = 0.$$

Hence,

$$\mu'_m = \mathbf{E}(X^m) = 0.$$

It follows that

$$X = 0, \qquad M_X(t) = 1.$$

Corollary 8.26 *There are functions that cannot be represented as m.g.f.*

Theorem 8.27 *Let X be a random variable with the m.g.f. $M(t)$. Assume $\exists \varepsilon : \varepsilon > 0$ and $M(t)$ is defined $\forall t \in (-\varepsilon, \varepsilon)$. The distribution of X is said to be uniquely defined by this function $M(t)$.*

Proof. We show only that $\mathbf{E}(h(X))$ is uniquely defined by $M(t)$ for any function h represented as Taylor series

$$h(x) = h(0) + h'(0)x + h''(0)\frac{x^2}{2!} + ... + \frac{d^k h}{dx^k}(0)\frac{x^k}{k!} + ...$$

This gives

$$\mathbf{E}(h(X)) = \mathbf{E}\Big(h(0) + h'(0)X + h''(0)\frac{X^2}{2!} + ... + \frac{d^k h}{dx^k}(0)\frac{X^k}{k!} + ...\Big)$$
$$= h(0) + h'(0)\mathbf{E}(X) + h''(0)\frac{\mathbf{E}(X^2)}{2!} + ... + \frac{d^k h}{dx^k}(0)\frac{\mathbf{E}(X^k)}{k!} + ...$$

Therefore, for a given function h, $\mathbf{E}(h(X))$ is uniquely defined by $M(t)$.

Example 8.28 Let X be a random variable with the m.g.f. $M_X(t)$ such that

$$M_X(t) = 1$$

for all $t \in (-\varepsilon, \varepsilon)$, where $\varepsilon > 0$. Then $X = 0$.

Proof. Consider a constant $Z = 0$ as a discrete random variable with frequency

$$p_Z(0) = \mathbf{P}(Z = 0) = 1.$$

It has the m.g.f.

$$M_Z(t) = \mathbf{E}e^{0 \cdot t} = 1.$$

By the theorem, it follows that X has the same distribution as Z, i.e., it is a discrete random variable with frequency

$$p_X(0) = \mathbf{P}(X = 0) = 1.$$

Example 8.29 Let X be a random variable with the m.g.f. $M_X(t)$ such that

$$M_X(t) = e^{at}$$

for all $t \in (-\varepsilon, \varepsilon)$, where $\varepsilon > 0$, $a \in \mathbf{R}$. Then $X = a$.

Proof. Consider a constant $Z = a$ as a discrete random variable with frequency

$$p_Z(a) = \mathbf{P}(Z = a) = 1.$$

It has the m.g.f.

$$M_Z(t) = \mathbf{E}e^{a \cdot t} = e^{at}.$$

Hence, X has the same distribution as Z, i.e. it is a discrete random variable with frequency

$$p_X(a) = \mathbf{P}(X = a) = 1.$$

Example 8.30 Let X be a random variable with the m.g.f. $M_X(t)$ such that

$$M_X(t) = \alpha + pe^t,$$

where $\alpha \in \mathbf{R}$, $p \in [0, 1]$. Then X is a Bernoulli random variable with the parameter p. In addition, $\alpha = 1 - p$.

Proof. Since $M(0) = 1$, it follows that $\alpha + p = 1$. We have that $M_X(t)$ is the m.g.f. for a Bernoulli random variable.

Example 8.31 Let X, Y be independent random variables with the m.g.f.

$$M_X(t) = M_Y(t) = \sqrt{\frac{\lambda}{\lambda - t}}$$

for all $t \in (-\varepsilon, \varepsilon)$, where $\varepsilon \in (0, \lambda)$, $\lambda > 0$. Then $Z = X + Y$ has the exponential distribution with parameter α.

Proof. We have that

$$M_Z(t) = M_Y(t) M_X(t) = \frac{\lambda}{\lambda - t}.$$

It is a m.g.f. for an exponential distribution (see example above).

Problems for Week 8

Problem 8.1 *Assume that* Y *has exponential distribution* $Exp(3)$. *Assume that* $X = 2^Y$. *Find* $\mathbf{E}X$.

Problem 8.2 *Assume that* X *has moment generating function*

$$M_X(t) = \frac{e^t + 1}{2 - t}.$$

Find $\mathbf{E}X$ *and* $\mathbf{E}X^2$.

Problem 8.3 *Assume that* X *has moment generating function*

$$M_X(t) = \frac{e^{3t} + e^{4t}}{2 - t}.$$

Find $\mathbf{E}X$.

Problem 8.4 *Let* X_1, X_2, *and* X_3 *be independent random variables with the same density*

$$f(x) = \begin{cases} 2e^{-2x}, & x \geq 0 \\ 0, & \text{otherwise.} \end{cases}$$

Let $Y = \min(X_1, X_2, X_3)$, *where* $X_i \sim Exp(2)$.

(a) Find the distribution of Y *and the moment generating function for* Y.
(b) Use the moment generating function to find $\mathbf{E}(Y)$ *and* $\mathbf{E}(Y^2)$.

Problem 8.5 *Assume that* X *has the density*

$$f(x) = \begin{cases} ce^{-2x}, & x \geq 1 \\ 0, & \text{otherwise,} \end{cases}$$

where $c > 0$ *is a constant.*

(a) Calculate the m.g.f. $M_Y(t)$.
(b) Find $\mathbf{E}(X)$ *using* $M(t)$.
(c) Prove that $X = a + Y$, *where* $a \in \mathbf{R}$, Y *has an exponential distribution.*

Week 9. Analysis of Some Important Distributions

In this chapter, we will apply the moment generating function to analyze some important distributions, including Normal, Gamma, Chi-square, and Poisson distributions. We will be using the concepts established above.

In particular, we will be using the following rule. Let a, b be some real numbers with $b \neq 0$. If X has the density function f_X, then $Y = a + bX$ has the density function

$$f_Y(y) = |b|^{-1} f_X\left(\frac{y-a}{b}\right),$$

and

$$\mathbf{E}(Y) = a + b\mathbf{E}(X), \quad \mathrm{Var}\,(Y) = b^2 \mathrm{Var}\,(X).$$

9.1 Normal (Gaussian) distribution

Definition 9.1 *We say that X has the standard normal distribution if it has the density*

$$\phi(x) = \frac{1}{\sqrt{2\pi}} \exp\left(-\frac{x^2}{2}\right).$$

We call this distribution $N(0,1)$.

Theorem 9.2 $\mathbf{E}X = 0$.

Proof. Since X has a symmetric density $\phi(x) = \phi(-x)$, it follows that

$$\begin{aligned}\mathbf{E}X = \int_{-\infty}^{\infty} x\phi(x)dx &= \int_{-\infty}^{0} x\phi(x)dx + \int_{0}^{\infty} x\phi(x)dx \\ &= -\int_{0}^{\infty} z\phi(-z)dz + \int_{0}^{\infty} x\phi(x)dx \\ &= -\int_{0}^{\infty} z\phi(z)dz + \int_{0}^{\infty} x\phi(x)dx = 0.\end{aligned}$$

Theorem 9.3 *The moment generating function for X is*

$$M(t) = \mathbf{E}e^{tX} = \exp\left(\frac{t^2}{2}\right).$$

Proof. We have

$$M(t) = \mathbf{E}e^{tX} = \int_{-\infty}^{+\infty} e^{tx}\phi(x)dx = \frac{1}{\sqrt{2\pi}}\int_{-\infty}^{+\infty} e^{tx}\exp\left(-\frac{x^2}{2}\right)dx.$$

Clearly,

$$tx - \frac{x^2}{2} = -\frac{(x-t)^2}{2} + \frac{t^2}{2}.$$

Hence

$$\begin{aligned} M(t) &= \frac{1}{\sqrt{2\pi}}\int_{-\infty}^{+\infty} \exp\left(-\frac{(x-t)^2}{2} + \frac{t^2}{2}\right)dx \\ &= \exp\left(\frac{t^2}{2}\right)\frac{1}{\sqrt{2\pi}}\int_{-\infty}^{+\infty}\exp\left(-\frac{(x-t)^2}{2}\right)dx. \end{aligned}$$

With the change of variables $y = x - t$, we obtain

$$\frac{1}{\sqrt{2\pi}}\int_{-\infty}^{+\infty}\exp\left(-\frac{(x-t)^2}{2}\right)dx = \frac{1}{\sqrt{2\pi}}\int_{-\infty}^{+\infty}\exp\left(-\frac{y^2}{2}\right)dy = 1.$$

We used here the fact that $\frac{1}{\sqrt{2\pi}}\exp\left(-\frac{y^2}{2}\right)$ is the density for $N(0,1)$. Hence

$$M(t) = \exp\left(\frac{t^2}{2}\right).$$

Theorem 9.4 $\mathbf{E}(X^2) = 1$, $\mathrm{Var}\,(X) = 1$.

Proof. We have that

$$\frac{dM}{dt}(t) = \exp\left(\frac{t^2}{2}\right)\frac{d}{dt}\left(\frac{t^2}{2}\right) = t\exp\left(\frac{t^2}{2}\right) = tM(t).$$

Hence

$$\frac{d^2M}{dt^2}(t) = \frac{d}{dt}(tM(t)) = M(t) + t\frac{dM}{dt}(t) = M(t) + t^2M(t).$$

Therefore, $\mathbf{E}(X)^2 = \frac{d^2M}{dt^2}(0) = M(0) = 1$.

Definition 9.5 *Let $X \sim N(0,1)$ and $Y = \mu + \sigma X$. We say that Y has the normal (Gaussian) distribution with the parameters μ and σ, i.e., that Y has the distribution $N(\mu, \sigma^2)$.*

Theorem 9.6 *Let Y has the distribution $N(\mu, \sigma^2)$. Then the following holds.*

(a) Y has the density

$$f_Y(x) = \sigma^{-1}\phi\left(\frac{x-\mu}{\sigma}\right) = \frac{1}{\sigma\sqrt{2\pi}}\exp\left(-\frac{(x-\mu)^2}{2\sigma^2}\right).$$

(b) The expected value and variance value of Y are

$$\mathbf{E}(Y) = \mu, \quad \mathrm{Var}\,(Y) = \sigma^2.$$

(c) The m.g.f. for Y is

$$M_Y(t) = e^{\mu t}e^{\frac{b^2t^2}{2}}.$$

Theorem 9.7 *Let $X \sim N(\mu, \sigma^2)$ and $Z = (X - \mu)/\sigma$. Then $Z \sim N(0, 1)$.*

Theorem 9.8 *Let X, Y be independent Gaussian random variables from $N(\mu_X, \sigma_X^2)$ and $N(\mu_Y, \sigma_Y^2)$. Then $Z = X + Y$ is a Gaussian random variable $N(\mu_Z, \sigma_Z^2)$, where*

$$\mathbf{E}Z = \mu_Z = \mu_X + \mu_Y,$$
$$\mathrm{Var}\,Z = \sigma_Z^2 = \sigma_X^2 + \sigma_Y^2.$$

Proof. The equations for $\mathbf{E}(Z)$ and $\mathrm{Var}\,Z$ follow from known properties of expectation and variance for a sum. Let us find the m.g.f. $M_Z(t)$ for z. We have that

$$M_Z(t) = M_X(t)M_Y(t) = e^{\mu_X t}e^{\frac{\sigma_X^2 t^2}{2}}e^{\mu_Y t}e^{\frac{\sigma_Y^2 t^2}{2}}.$$

It gives

$$M_Z(t) = M_X(t)M_Y(t) = e^{(\mu_X+\mu_Y)t}e^{\frac{(\sigma_X^2+\sigma_Y^2)t^2}{2}} = e^{\mu_Z t \frac{\sigma_Z^2 t^2}{2}}.$$

We know that a m.g.f. defines the distribution uniquely. Since it is a m.g.f. for a normal distribution $N(\mu_Z, \sigma_Z^2)$, it follows that Z has this distribution.

9.2 Bivariate normal density

Let us consider two dimensional case. We say that a pair of random variables X and Y has a bivariate normal density if their joint density is given as

$$f(x,y) = \frac{1}{2\pi\sigma_X\sigma_Y\sqrt{1-\rho^2}}$$
$$\times \exp\left(-\frac{1}{2(1-\rho^2)}\left[\frac{(x-\mu_X)^2}{\sigma_X^2} + \frac{(y-\mu_Y)^2}{\sigma_Y^2} - \frac{2\rho(x-\mu_X)(y-\mu_Y)}{\sigma_X\sigma_Y}\right]\right),$$

where $x, y \in \mathbf{R}$. The density depends on 5 parameters:

$$\mu_X, \mu_Y \in \mathbf{R}, \quad \sigma_X > 0, \quad \sigma_Y > 0, \quad \rho \in (-1, 1).$$

In this case, we also say that the vector (X, Y) has Gaussian distribution, or normal distribution, or that (X, Y) is a Gaussian random vector.

Theorem 9.9 *The marginal densities for the bivariate normal density are* $N(\mu_X, \sigma_X^2)$ *and* $N(\mu_Y, \sigma_Y^2)$.

Remark 9.10 *It may happen that* X *and* Y *are both Gaussian random variables, but the vector* (X, Y) *is not Gaussian.*

Corollary 9.11

$$E(X) = \mu_X, \quad \mathrm{Var}\,(X) = \sigma_X^2,$$
$$\mathbf{E}(Y) = \mu_Y, \quad \mathrm{Var}\,(Y) = \sigma_Y^2.$$

Theorem 9.12 *Let* X *and* Y *be independent Gaussian random variables,* $X \sim N(\mu_X, \sigma_X^2)$ *and* $Y \sim N(\mu_Y, \sigma_Y^2)$. *Then*

(a) $\mathrm{corr}(X, Y) = 0$,
(b) the joint density for (X, Y) *is given as above with* $\rho = 0$, *i.e.,*

$$f(x, y) = \frac{1}{2\pi\sigma_X\sigma_Y} \exp\left(-\frac{1}{2}\left[\frac{(x-\mu_X)^2}{\sigma_X^2} + \frac{(y-\mu_Y)^2}{\sigma_Y^2}\right]\right), \quad x, y \in \mathbf{R}.$$

Proof. From the independence, we obtain that $\mathrm{corr}(X, Y) = 0$. Further, observe that the joint density

$$f(x, y) = f_X(x) f_Y(y),$$

where

$$f_X(x) = \frac{1}{\sigma_X\sqrt{2\pi}} \exp\left(-\frac{(x-\mu_X)^2}{2\sigma_X^2}\right),$$
$$f_Y(y) = \frac{1}{\sigma_Y\sqrt{2\pi}} \exp\left(-\frac{(y-\mu_Y)^2}{2\sigma_Y^2}\right).$$

This completes the proof.

Theorem 9.13

$$\rho = \mathrm{corr}(X, Y).$$

Corollary 9.14 *Let (X, Y) be a Gaussian random vector such that*

$$\operatorname{corr}(X, Y) = 0.$$

Then X and Y are independent.

This is a very important feature of Gaussian distributions.

Example 9.15 Assume that the joint density for (X, Y) is

$$f(x, y) = ce^{(-x^2-4y^2-xy)},$$

where $c > 0$ is a constant. Let us find the marginal densities, the corresponding expectations, variances and the correlation.

Solution. From the equation for f, we see that this is a bivariate normal density such that

$$\frac{1}{2(1-\rho^2)} \frac{(x-\mu_X)^2}{\sigma_X^2} = x^2,$$

$$\frac{1}{2(1-\rho^2)} \frac{(y-\mu_Y)^2}{\sigma_Y^2} = 4y^2,$$

$$\frac{1}{2(1-\rho^2)} \frac{2\rho(x-\mu_X)(y-\mu_Y)}{\sigma_X\sigma_Y} = -xy.$$

It gives that $\mu_X = \mu_Y = 0$ and

$$\frac{1}{2(1-\rho^2)\sigma_X^2} = 1, \quad \frac{1}{2(1-\rho^2)\sigma_Y^2} = 4, \quad \frac{2\rho}{2(1-\rho^2)\sigma_X\sigma_Y} = -1.$$

From the first two equations, we obtain that

$$2(1-\rho^2)\sigma_X^2 \cdot 2(1-\rho^2)\sigma_Y^2 = 1/4.$$

Hence

$$2(1-\rho^2)\sigma_X\sigma_Y = 1/2.$$

It gives

$$\frac{2\rho}{2(1-\rho^2)\sigma_X\sigma_Y} = 4\rho, \quad 4\rho = -1.$$

Hence $\rho = -1/4$. Finally, we obtain

$$\sigma_X = \sqrt{\frac{1}{2(1-\rho^2)}} = 0.73, \quad \sigma_Y = \sqrt{\frac{1}{4 \cdot 2(1-\rho^2)}} = 0.36,$$

and

$$X \sim N(0, 0.73^2), \quad Y \sim N(0, 0.36^2), \quad \operatorname{corr}(X, Y) = -1/4.$$

Theorem 9.16 *Let (X, Y) be a Gaussian random vector, i.e. X, Y such that the distribution of (X, Y) is bivariate Gaussian. Assume that the distributions for X and Y are $N(\mu_X, \sigma_X^2)$ and $N(\mu_Y, \sigma_Y^2)$ respectively. Then $Z = X + Y$ is a Gaussian random variable $N(\mu_Z, \sigma_Z^2)$, where*

$$\mu_Z = \mathbf{E}Z = \mu_X + \mu_Y,$$
$$\sigma_Z^2 = \operatorname{Var} Z = \sigma_X^2 + 2\operatorname{corr}(X, Y)\sigma_X\sigma_Y + \sigma_Y^2.$$

Theorem 9.17 *Let (X, Y) be a Gaussian random vector,*

$$X \sim N(\mu_X, \sigma_X^2), \quad Y \sim N(\mu_Y, \sigma_Y^2), \quad \operatorname{corr}(X, Y) = \rho,$$

and

$$Z = Y - \rho\frac{\sigma_Y}{\sigma_X}X.$$

Then X and Z are independent, $\operatorname{Cov}(X, Z) = 0$ *and* $\operatorname{corr}(X, Z) = 0$.

Proof.

$$\operatorname{Cov}(X, Z) = \operatorname{Cov}(X, Y) - \operatorname{Cov}(X, \rho\frac{\sigma_Y}{\sigma_X}X) = \rho\sigma_X\sigma_Y - \rho\frac{\sigma_Y}{\sigma_X}\sigma_X^2 = 0.$$

Corollary 9.18 *(The best estimate for a component of a Gaussian random vector):*

$$\mathbf{E}(Y|X) = \mu_Y + \rho\frac{\sigma_Y}{\sigma_X}(X - \mu_X).$$

Proof. Let $Z = Y - \frac{\rho\sigma_Y}{\sigma_X}X$. We have that

$$Y = Z + \rho\frac{\sigma_Y}{\sigma_X}X.$$

Hence

$$\mathbf{E}(Y|X) = \mathbf{E}(Z|X) + \rho\frac{\sigma_Y}{\sigma_X}X.$$

Since Z is independent of X, it follows that

$$\mathbf{E}(Z|X) = \mathbf{E}Z.$$

Further,

$$\mathbf{E}Z = \mathbf{E}Y - \mathbf{E}\rho\frac{\sigma_Y}{\sigma_X}X = \mu_Y - \rho\frac{\sigma_Y}{\sigma_X}\mu_X.$$

Finally, we obtain

$$\mathbf{E}(Y|X) = \mu_Y - \rho\frac{\sigma_Y}{\sigma_X}\mu_X + \rho\frac{\sigma_Y}{\sigma_X}X = \mu_Y + \rho\frac{\sigma_Y}{\sigma_X}(X - \mu_X).$$

Example 9.19 Let (X, Y) be a Gaussian random vector,

$$X \sim N(1.1, 4), \quad Y \sim N(-1, 9), \quad \text{corr}(X, Y) = -1/2.$$

We have $\sigma_X = 2$, $\sigma_Y = 3$,

$$\mathbf{E}(Y|X) = -1 + \left(-\frac{1}{2}\right)\frac{3}{2}(X - 1.1).$$

Assume that we observe $X = 1$. Then the best estimate of Y is

$$\mathbf{E}(Y|X = 1) = -1 + \left(-\frac{1}{2}\right)\frac{3}{2}(1 - 1.1) = -1 + 3/4 \cdot 0.1 = -0.9250.$$

Example 9.20 Let (X, Y) be a Gaussian random vector,

$$X \sim N(1.1, 4), \quad Y \sim N(-1, 9), \quad \text{corr}(X, Y) = 1/2.$$

Then

$$\mathbf{E}(Y|X) = -1 + \frac{1}{2} \cdot \frac{3}{2}(X - 1.1).$$

Assume that we observe $X = 1$. Then the best estimate of Y is

$$\mathbf{E}(Y|X = 1) = -1 + \frac{1}{2}\frac{3}{2}(1 - 1.1) = -1 - 3/4 \cdot 0.1 = -1.0750.$$

Example 9.21 Let (X, Y) be a Gaussian random vector,

$$X \sim N(1.1, 4), \quad Y \sim N(-1, 9), \quad \text{corr}(X, Y) = 0.$$

Then

$$\mathbf{E}(Y|X) = -1 + 0 \cdot \frac{3}{2}(X - 1.1) = -1.$$

In this case, Y and X are independent, hence

$$\mathbf{E}(Y|X) = \mathbf{E}(Y) = -1.$$

9.3 Chi-squared distribution

Definition 9.22 *Let X be a random variables with standard normal distribution $N(0, 1)$. The distribution of X^2 has χ^2-distribution (Chi-squared distribution) with one degree of freedom (or χ^2_1-distribution).*

Definition 9.23 *Let $X_1, ..., X_n$ be independent Gaussian random variables with standard normal distribution $N(0, 1)$, and*

$$Z = X_1^2 + X_2^2 + ... + X_n^2.$$

We say that Z has χ^2-distribution with n degrees of freedom (or χ^2_n-distribution).

Theorem 9.24 *Moment generating function for χ^2-distribution with n degrees of freedom is*

$$M(t) = (1-2t)^{-n/2}.$$

It is defined for $t < \frac{1}{2}$.

Proof. Let us first find the moment generating function $M_i(t)$ for χ_1^2-distribution (or χ^2-distribution with 1 degree of freedom), i.e., where $Z = X_i^2$, $X_i \sim N(0,1)$. We have

$$M_i(t) = \mathbf{E}e^{tZ} = e^{tX^2}.$$

Hence

$$\begin{aligned} M_i(t) &= \frac{1}{\sqrt{2\pi}} \int_{-\infty}^{\infty} e^{tx^2} \exp\left(-\frac{x^2}{2}\right) dx \\ &= \frac{1}{\sqrt{2\pi}} \int_{-\infty}^{\infty} \exp\left(-\frac{x^2 - 2tx^2}{2}\right) dx. \end{aligned}$$

Let $\sigma = \sigma(t)$ be defined such that

$$\frac{(1-2t)x^2}{2} = \frac{x^2}{2\sigma(t)^2}.$$

It gives

$$\sigma(t)^2 = \frac{1}{1-2t}.$$

Thus

$$M_i(t) = \sigma(t) \cdot \frac{1}{\sigma(t)\sqrt{2\pi}} \int_{-\infty}^{\infty} \exp\left(-\frac{x^2}{2\sigma(t)^2}\right) dx = \sigma(t) \cdot 1 = (1-2t)^{-1/2}.$$

Finally, we obtain that the m.g.f. for $Z = X_1^2 + X_2^2 + ... + X_n^2$ is

$$M(t) = M_1(t)M_2(t) \times \cdots \times M_n(t) = (1-2t)^{-n/2}.$$

Theorem 9.25 *Let X has χ_n^2-distribution. Then*

$$\mathbf{E}(X) = n.$$

Proof. The m.g.f. is

$$M(t) = (1-2t)^{-n/2}.$$

Hence

$$M'(t) = 2 \cdot \frac{n}{2}(1-2t)^{-n/2-1}.$$

It gives $\mathbf{E}(X) = M'(0) = n$.

Alternatively, we can simply observe that

$$\mathbf{E}X = \mathbf{E}(X_1^2 + ... + X_n^2) = \mathbf{E}X_1^2 + ... + \mathbf{E}X_n^2 = n.$$

We use tables for calculating probabilities for χ^2-distributions.

Table A3 shows the value c such that $\mathbf{P}(Z > c) = p$, where $p = 0.995, 0.99, 0.975, ...$, Z has a χ^2-distribution.

degrees of freedom (v)	..	$p = 0.95$	....	....
..	...	...	....	...
2	...	0.103	...	...

Example 9.26 Let X_1 and X_2 be independent random variables, $X_i \sim N(0,1)$ and $Z = X_1^2 + X_2^2$. The distribution of Z is χ^2 with 2 degrees of freedom (χ_2^2). Let us estimate $\mathbf{P}(Z < 0.103)$. By table A3, we obtain that $\mathbf{P}(Z \le 0.103) = 0.05$.

Example 9.27 Let X_1 and X_2 be independent random variables where $X_i \sim N(0, 0.25)$ and $Z = X_1^2 + X_2^2$. Let us estimate $\mathbf{P}(Z < 0.0255)$.

Solution. It is the distribution $N(0, \sigma^2)$ with $\sigma = 0.5$. We have that

$$X_i = \sigma Y_i = 0.5 Y_i,$$

where $Y_i \sim N(0,1)$. Hence

$$Z = X_1^2 + X_2^2 = \sigma^2(Y_1^2 + Y_2^2) = \frac{1}{4}(Y_1^2 + Y_2^2).$$

We know that $U = Y_1^2 + Y_2^2 \sim \chi_2^2$. Therefore, $4Z$ has the distribution χ_2^2 (i.e., χ^2 distribution with two degrees of freedom). We have

$$\mathbf{P}(Z < 0.02575) = \mathbf{P}(4Z < 0.103) = \mathbf{P}(U < 0.103).$$

By table A3, we obtain

$$\mathbf{P}(Z < 0.02575) = \mathbf{P}(U \le 0.103) = 0.05.$$

9.4 Poisson distribution

Let $\lambda > 0$ be given. We repeat that a discrete random variable X has a Poisson frequency function with parameter λ if

$$\mathbf{P}(X = k) = \frac{\lambda^k}{k!} e^{-\lambda}, \quad k \ge 0.$$

Theorem 9.28 *For the Poisson distribution, the moment generation function is*

$$M(t) = e^{-\lambda} e^{\lambda e^t}.$$

Proof.

$$M(t) = \mathbf{E}e^{tX} = \sum_{k=0}^{\infty} e^{tk}\mathbf{P}(X = k) = e^{-\lambda} \sum_{k=0}^{\infty} e^{tk} \frac{\lambda^k}{k!} = e^{-\lambda} \sum_{k=0}^{\infty} \frac{(\lambda e^t)^k}{k!} = e^{-\lambda} e^{\lambda e^t}.$$

We use the Taylor series for e^z with $z = \lambda e^t$.

Theorem 9.29 *For Poisson distribution,* $\mathbf{E}X = \lambda$.

Proof. It suffices to observe that

$$\frac{dM}{dt}(0) = e^{-\lambda} e^{\lambda e^t} \lambda e^t \Big|_{t=0} = \lambda.$$

9.5 Gamma distribution

Definition 9.30 *For* $\alpha > 0$, *Gamma function is defined as*

$$\Gamma(\alpha) = \int_0^{\infty} u^{\alpha-1} e^{-u} du.$$

In particular, $\Gamma(n+1) = n!$ *for integers* $n > 0$.

Some useful integrals

We can use this definition as a rule of integration:

$$\int_0^{\infty} u^r e^{-u} du = \Gamma(r+1).$$

It can be also rewritten as

$$\int_0^{\infty} t^{\alpha} \lambda e^{-\lambda t} dt = \frac{\Gamma(\alpha+1)}{\lambda^{\alpha}}. \tag{9.4}$$

For an integer m, it gives

$$\int_0^{\infty} t^m \lambda e^{-\lambda t} dt = \frac{\Gamma(m+1)}{\lambda^m} = \frac{m!}{\lambda^m}.$$

Definition 9.31 *Gamma distribution is the distribution with the density*

$$f(x) = \begin{cases} \frac{\lambda^\alpha}{\Gamma(\alpha)} x^{\alpha-1} e^{-\lambda x}, & x \geq 0 \\ 0, & \text{otherwise.} \end{cases}$$

It depends on two parameters, α and λ.

If $\alpha = 1$ then it is an exponential density. The parameter α is called a shape parameter. Varying α changes the shape of the density.

Theorem 9.32 *For Gamma distribution, the m.g.f. is*

$$M(t) = \left(\frac{\lambda}{\lambda - t} \right)^\alpha.$$

It is defined for $t < \lambda$.

Proof.

$$M(t) = \int_0^\infty e^{tx} \frac{\lambda^\alpha}{\Gamma(\alpha)} x^{\alpha-1} e^{-\lambda x} dx = \frac{\lambda^\alpha}{\Gamma(\alpha)} \int_0^\infty x^{\alpha-1} e^{-(\lambda - t)x} dx.$$

Using integral (9.4), we obtain

$$M(t) = \frac{\lambda^\alpha}{\Gamma(\alpha)} \frac{\Gamma(\alpha)}{(t - \lambda)^\alpha} = \left(\frac{\lambda}{\lambda - t} \right)^\alpha.$$

Theorem 9.33 *Let X has Gamma distribution. Then*

$$\mathbf{E}(X) = \frac{\alpha}{\lambda}, \qquad \mathbf{E}(X^2) = \frac{\alpha(\alpha+1)}{\lambda^2}, \qquad \mathrm{Var}\,(X) = \frac{\alpha}{\lambda^2}.$$

Proof. The m.g.f. is

$$M(t) = \left(\frac{\lambda}{\lambda - t} \right)^\alpha.$$

Thus

$$M'(t) = \alpha \frac{\lambda^\alpha}{(\lambda - t)^{\alpha+1}}.$$

It gives

$$\mathbf{E}(X) = M'(0) = \frac{\alpha}{\lambda}.$$

Further, we obtain

$$M''(t) = \alpha \frac{d}{dt} \frac{\lambda^\alpha}{(\lambda - t)^{\alpha+1}} = \alpha(\alpha+1) \frac{\lambda^\alpha}{(\lambda - t)^{\alpha+2}}.$$

Therefore

$$\mathbf{E}(X^2) = M''(0) = \frac{\alpha(\alpha+1)}{\lambda^2}.$$

Finally,

$$\mathrm{Var}\,(X) = \frac{\alpha(\alpha+1)}{\lambda^2} - \frac{\alpha^2}{\lambda^2} = \frac{\alpha}{\lambda^2}.$$

Example 9.34 Let X, Y be independent random variables with the m.g.f.

$$M_X(t) = M_Y(t) = \sqrt{\frac{\lambda}{\lambda - t}},$$

for all $t < \lambda$ and $\lambda > 0$. This is a Gamma distribution with $\alpha = 1/2$ and λ. Hence

$$\mathbf{E}(X) = \mathbf{E}(Y) = \frac{\alpha}{\lambda} = \frac{1}{2\lambda}.$$

Under the assumptions of the above example, $Z = X + Y$ has the exponential distribution with parameter λ and with the m.g.f.

$$M_Z(t) = M_Y(t)M_X(t) = \frac{\lambda}{\lambda - t}.$$

Hence $\mathbf{E}Z = \frac{1}{\lambda}$. It is agreed with the fact that $\mathbf{E}(Z) = \mathbf{E}(X) + \mathbf{E}(Y)$.

χ^2 *distribution as a special case of Gamma distribution*

Let X has χ^2 distribution with n degrees of freedom. Its m.g.f. is

$$M(t) = \frac{1}{(1-2t)^{n/2}} = \frac{(1/2)^{n/2}}{(1/2-t)^{n/2}} = \frac{\lambda^\alpha}{(\lambda - t)^\alpha},$$

where $\lambda = 1/2$ and $\alpha = n/2$. It is the m.g.f. for Gamma distribution. Hence χ^2 is a special case of Gamma distribution with $\lambda = 1/2$, $\alpha = n/2$.

Corollary 9.35 *Let $X \sim \chi^2_n$, let $\lambda = 1/2$, and let $\alpha = n/2$. Then*

$$\mathbf{E}(X) = \frac{\alpha}{\lambda} = n, \quad \mathbf{E}(X^2) = \frac{\alpha(\alpha+1)}{\lambda^2}, \quad \mathrm{Var}\,(X) = \frac{\alpha}{\lambda^2} = 2n,$$

and the density for X is

$$g(t) = \begin{cases} \frac{\lambda^\alpha}{\Gamma(\alpha)} t^{\alpha-1} e^{-\lambda t}, & x \geq 0 \\ 0, & \textit{otherwise}. \end{cases}$$

Example 9.36 Let us apply the theory for calculation of $\mathbf{E}(Y^4)$, where $Y \sim N(0,1)$. We have $Y^4 = X^2$, where $X = Y^2$, $X \sim \chi^2_1$. Hence

$$\mathbf{E}(Y^4) = \mathbf{E}(X^2) = \frac{\alpha(\alpha+1)}{\lambda^2}, \quad \lambda = 1/2, \alpha = n/2 = 1/2.$$

It gives

$$\mathbf{E}(Y^4) = \mathbf{E}(X^2) = \frac{3/4}{1/4} = 3.$$

9.6 Some other distributions

Lognormal distribution

We say that X has a lognormal distribution if $X = e^Y$, where $Y \sim N(\mu, \sigma^2)$. It has the density

$$f(x) = \frac{1}{x\sigma\sqrt{2\pi}} \exp\left(-\frac{(\log x - \mu)^2}{2\sigma^2}\right), \quad x > 0.$$

It can be rewritten as $X \sim LN(\mu, \sigma^2)$.

We have that $M_Y(t) = e^{\mu t} e^{\sigma^2 t^2/2}$. Hence

$$\mathbf{E}X == \mathbf{E}e^Y = e^{\mu + \frac{1}{2}\sigma^2}$$

and

$$\mathbf{E}X^2 = \mathbf{E}e^{2Y} = M_Y(2) = e^{2\mu + 4\frac{1}{2}\sigma^2} = e^{2\mu + 2\sigma^2}.$$

Hence

$$\operatorname{Var} X = \mathbf{E}X^2 - (\mathbf{E}X)^2 = e^{2\mu+2\sigma^2} - e^{2\mu + 2\frac{1}{2}\sigma^2} = e^{2\mu+\sigma^2}\left(e^{\sigma^2} - 1\right).$$

We can also obtain

$$\mathbf{E}X^k = \mathbf{E}e^{kY} = M_Y(k) = e^{k\mu + k^2\frac{1}{2}\sigma^2}.$$

Weibull distribution

We say that X has the Weibull distribution if it has the c.d.f.

$$F(x) = 1 - \exp(-cx^\gamma), \quad x \geq 0.$$

Here $\gamma > 0$ and $c > 0$. It can be rewritten as $X \sim W(c, \gamma)$. Its p.d.f. is

$$f(x) = c\gamma x^{\gamma-1} \exp(-cx^\gamma), \quad x > 0.$$

Let us show that for $X \sim W(c, \gamma)$,

$$\mathbf{E}(X) = \frac{\Gamma(1 + 1/\gamma)}{c^{1/\gamma}}.$$

This is because

$$\mathbf{E}X = \int_0^{+\infty} x f(x) dx = \int_0^{+\infty} x c\gamma x^{\gamma-1} \exp(-cx^\gamma) dx.$$

Substituting $u = cx^\gamma$ gives

$$du = c\gamma x^{\gamma-1} dx, \quad x = (u/c)^{1/\gamma}.$$

Hence

$$\mathbf{E}X = \int_0^{+\infty} (u/c)^{1/\gamma} e^{-u} du.$$

Using integration formula (9.4), we obtain that

$$\mathbf{E}X = \int_0^{+\infty} (u/c)^{1/\gamma} e^{-u} du = c^{-1/\gamma}\Gamma(1 + 1/\gamma).$$

Generalized Pareto distribution

If X has a generalized Pareto distribution (or 3-parameter Pareto distribution), then its density is

$$f(x) = \frac{\Gamma(\alpha+k)\lambda^{\alpha}x^{k-1}}{\Gamma(\alpha)\Gamma(k)(\lambda+x)^{\alpha+k}}, \quad x \geq 0.$$

Let $\alpha > 1$. Let us find $\mathbf{E}X$.

$$\mathbf{E}X = \int_0^{+\infty} xf(x)dx = \int_0^{+\infty} x\frac{\alpha\lambda^{\alpha}}{(\lambda+x)^{\alpha+1}}dx.$$

Let $k = 2$. We have

$$x\frac{\alpha\lambda^{\alpha}}{(\lambda+x)^{\alpha+1}} = \lambda\frac{\alpha\lambda^{\alpha-1}x^{k-1}}{(\lambda+x)^{\alpha-1+k}} = \lambda\frac{\alpha\lambda^{\alpha}x^{k-1}}{(\lambda+x)^{\alpha-1+k}}.$$

Note that

$$\Gamma(2) = 1 \cdot \Gamma(1) = 1,$$

$$\Gamma(\alpha+1) = \alpha\Gamma(\alpha) = \alpha(\alpha-1)\Gamma(\alpha).$$

Hence the density for the generalized Pareto distribution with the parameters $\alpha - 1$ and $k = 2$ is

$$f_{\alpha-1}(x) = \frac{\Gamma(\alpha-1+k)\lambda^{\alpha-1}x^{k-1}}{\Gamma(\alpha-1)\Gamma(k)(\lambda+x)^{\alpha-1+k}} = \frac{\Gamma(\alpha+1)\lambda^{\alpha-1}x^{k-1}}{\Gamma(\alpha-1)\Gamma(2)(\lambda+x)^{\alpha-1+k}}.$$

Hence

$$f_{\alpha-1}(x) = \alpha(\alpha-1)\frac{\lambda^{\alpha-1}x^{k-1}}{(\lambda+x)^{\alpha-1+k}},$$

$$x\frac{\alpha\lambda^{\alpha}}{(\lambda+x)^{\alpha+1}} = \frac{\lambda}{(1-\alpha)}f_{\alpha-1}(x).$$

Clearly,

$$\int_0^{\infty} f_{\alpha-1}(x)dx = 1.$$

It follows that

$$\mathbf{E}X = \int_0^{+\infty} x\frac{\alpha\lambda^{\alpha}}{(\lambda+x)^{\alpha+1}}dx = \int_0^{\infty} \frac{\lambda}{(1-\alpha)}f_{\alpha-1}(x)dx = \frac{\lambda}{(1-\alpha)}.$$

Burr distribution

We say that X has the Burr distribution if it has the c.d.f.

$$F(x) = 1 - \left(\frac{\lambda}{\lambda + x^\gamma}\right)^\alpha, \quad x \geq 0.$$

Here $\gamma > 0$, $\alpha > 0$ and $\lambda > 0$.

For the Burr distribution,

$$\mathbf{E}(X) = \frac{\lambda^{1/\gamma}\Gamma(\alpha - 1/\gamma)\Gamma(1 + 1/\gamma)}{\Gamma(\alpha)}.$$

Let us simplify it for $\gamma = 1$ (i..e, for the Pareto distribution): we have that

$$(\alpha - 1)\Gamma(\alpha - 1) = \Gamma(\alpha), \quad \Gamma(2) = 1.$$

Hence

$$\mathbf{E}(X) = \frac{\lambda\Gamma(\alpha - 1)\Gamma(2)}{\Gamma(\alpha)} = \frac{\lambda}{\alpha - 1}.$$

Problems for Week 9

Problem 9.1 *Let* $X \sim N(7,3)$, $Y \sim N(5,2)$, $\text{corr}(X,Y) = -0.2$.

(a) Find $\mathbf{E}(X|Y)$.

(b) Assume that an observer knows that $Y = 0.3$. *Suggest the best estimate / forecast of* X *given this information. Use (a).*

(c) Find the distribution for $Z = 2 + 0.5X + 0.1Y$. *Find* $\mathbf{P}(Z < 5)$.

Use known properties of sums and conditional expectations of pairs of Gaussian random variables.

Problem 9.2 *Let* $X \sim N(0,1)$, $Y \sim N(0,1)$, *and* $Z = X^2 + Y^2$. *Find* 0.9*-quantile of* Z, *i.e., find* c *such that* $\mathbf{P}(Z \leq c) = 0.9$.

Problem 9.3 *Let* X_1, X_2, X_3 *be independent random variables.*

(a) Let $X_i \sim N(0,1)$, *and let* $U = X_1^2 + X_2^2 + X_3^2$. *Describe the distribution of* U. *Estimate* $\mathbf{P}(U < 6.25)$.

(b) Let $X_i \sim N(0,1)$, *and let* $W = 2(X_1^2 + X_2^2 + X_3^2)$. *Describe the distribution of* W. *Estimate* $\mathbf{P}(W < 0.432)$.

(c) Let $X_i \sim N(0,0.25)$, *and let* $V = (X_1^2 + X_2^2 + X_3^2)$. *Describe the distribution of* V. *Estimate* $\mathbf{P}(V < 1.95)$.

Hint: the distribution may be found in the list of known special distributions.

Problem 9.4 *For Problem 9.3c, find* $\mathbf{E}(V^{1/2})$.

Week 10. Limit Theorems

In this chapter, we discuss convergence of random variables, limit theorems, and applications to estimation of probabilities.

10.1 Classical limits for the sequences of real numbers

Definition 10.1 *Let $\{x_i\}$ be a sequence of real numbers and $x \in \mathbf{R}$ be a number. We say that*

$$x = \lim_{i \to +\infty} x_i,$$

or

$$x_i \to x \quad as \quad i \to +\infty,$$

if, for any $\varepsilon > 0$, there exists N such that $|x_i - x| \leq \varepsilon$ for any $i > N$.

If $x_i = 1/i$, then $x_i \to 0$ as $i \to +\infty$.

If

$$x_i = 1 + a + \frac{a^2}{2!} + \dots + \frac{a^i}{i!},$$

then $x_i \to e^a$ as $i \to +\infty$.

10.2 Types of limits for random variables

Let $\{X_i\}$ be a sequence of random variables and X be a random variable.

Definition 10.2 *We say that X_i converges to X in probability if, for any $\varepsilon > 0$,*

$$\mathbf{P}(|X_i - X| \geq \varepsilon) \to 0 \quad as \quad i \to +\infty.$$

Definition 10.3 *We say that X_i converges to X almost surely (or, equivalently, with probability 1), if*

$$\mathbf{P}\Big(|X_i - X| \to 0 \quad as \quad i \to +\infty\Big) = 1.$$

Definition 10.4 *We say that X_i converges to X in mean square, if*

$$\mathbf{E}(|X_i - X|^2) \to 0 \quad as \quad i \to +\infty.$$

10.3 The Law of Large Numbers

Let

$$\overline{X}_n = \frac{1}{n}(X_1 + ... + X_n),$$

where X_i are independent random variables. We assume that, for all i,

$$\mathbf{E}X_i = \mu, \quad \operatorname{Var} X_i = \sigma^2.$$

Theorem 10.5 *(The Law of Large Numbers)*

$$\mathbf{E}(|\overline{X}_n - \mu|^2) \to 0 \quad as \quad n \to +\infty.$$

In other words, $\overline{X}_n$ converges in mean square to μ.

Proof. We have

$$\mathbf{E}\overline{X}_n = \mathbf{E}\left(\frac{1}{n}X_1\right) + ... + \mathbf{E}\left(\frac{1}{n}X_n\right) = \frac{1}{n}\mathbf{E}(X_1) + ... + \frac{1}{n}\mathbf{E}(X_n)$$
$$= \frac{n}{n}\mathbf{E}(X_i) = \mu,$$

and

$$\mathbf{E}(|\overline{X}_n - \mu|^2) = \operatorname{Var}\overline{X}_n = \operatorname{Var}\left(\frac{1}{n}X_1\right) + ... + \operatorname{Var}\left(\frac{1}{n}X_n\right)$$
$$= \frac{1}{n^2}\operatorname{Var}(X_1) + ... + \frac{1}{n^2}\operatorname{Var}(X_n) = \frac{n}{n^2}\operatorname{Var} X_i = \frac{1}{n}\sigma^2.$$

Corollary 10.6 *(The Weak Law of Large Numbers) For every $\varepsilon > 0$,*

$$\mathbf{P}(|\overline{X}_n - \mu| > \varepsilon) \to 0 \quad as \quad n \to +\infty.$$

Proof. By Chebyshev's Inequality,

$$\mathbf{P}(|\overline{X}_n - \mu| > \varepsilon) \le \frac{\operatorname{Var} \overline{X}_n}{\varepsilon^2}.$$

Hence

$$\mathbf{P}(|\overline{X}_n - \mu| > \varepsilon) \le \frac{\sigma^2}{n\varepsilon^2}.$$

Then the proof follows from the fact that

$$\frac{\sigma^2}{n\varepsilon^2} \to 0 \quad \text{as} \quad n \to +\infty.$$

Example 10.7 Let X_i be independent Bernoulli random variables where

$$X_i = \begin{cases} 1, & \text{with probability } p \\ 0, & \text{with probability } 1-p. \end{cases}$$

Then $\mathbf{E}X_i = p$ and

$$\overline{X}_n = \frac{1}{n}\sum_{i=1}^{n} X_i \to p \quad \text{as} \quad i \to +\infty$$

in mean square and in probability.

10.4 Convergence of random variables in distribution

Let $\{X_i\}$ be a sequence of random variables and X be a random variable. Define $F_{X_i}(x)$ and $F_X(x)$ as their corresponding c.d.f.

Definition 10.8 *We say that X_i converges to X in distribution if*

$$F_{X_i}(x) \to F_X(x) \quad \text{as} \quad i \to +\infty,$$

for any x where $F(x)$ is continuous.

Lemma 10.9 *X_i converges to X in distribution if and only if*

$$\mathbf{P}(X_i \in (a,b)) \to \mathbf{P}(X \in (a,b)) \quad \text{as} \quad i \to +\infty,$$

for any interval (a, b)

Example 10.10 Let $\mu \in \mathbf{R}$ and $\sigma > 0$ be some numbers. Let $\{\mu_i\}$ be a sequence such that $\mu_i \to \mu$ and $\{\sigma_i\}$ be a sequence such that $\sigma_i \to \sigma$, $\sigma_i > 0$ where

$$X_i \sim N(\mu_i, \sigma_i^2), \quad X \sim N(\mu, \sigma^2)$$

are Gaussian random variables. Then

$$X_i \to X \quad \text{in distribution.}$$

Note that this is true regardless of correlation or dependence of X_i and X.

Example 10.11 In the previous example, let $Y = 2\mu - X$. Then

$$X_i \to Y \quad \text{in distribution.}$$

Proof. Observe that $Y \sim N(\mu, \sigma^2)$, i.e., it has the same distribution as X.

Example 10.12 Let $X_1 = X_2 = ... = X_n$ and

$$X_i \sim N(0,1), \quad X \sim N(0,1).$$

Then

$$X_i \to X \quad \text{in distribution}$$

and at the same time,

$$X_i \to -X \quad \text{in distribution.}$$

Proof. Clearly,

$$\mathbf{P}(X_i \le a) = \mathbf{P}(X \le a))$$

for any i. Then $X_i \to X$ in distribution.

Further, let $Y = -X$. We have that $Y = aX$, where $a = -1$. In addition, $\mathbf{E}Y = a\mathbf{E}X = 0$, $\operatorname{Var} Y = a^2 \operatorname{Var} X = \operatorname{Var} X$. Hence $Y \sim N(0,1)$. It follows that

$$\mathbf{P}(X_i \le a) = \mathbf{P}(Y \le a)$$

for any i. Then $X_i \to Y = -X$ in distribution.

Example 10.13 Let $\lambda > 0$ be given. Let $\{\lambda_i\}$ be a sequence such that $\lambda_i \to \lambda$, $\lambda_i > 0$ and X_i be Poisson random variables with parameter λ_i. Then

$$X_i \to X \quad \text{as} \quad i \to +\infty \quad \text{in distribution.}$$

Proof. For any a,

$$\mathbf{P}(X_i \le a) = \sum_{k \le a} \mathbf{P}(X_i = k) = \sum_{k \le a} \frac{\lambda_i^k}{k!} e^{-\lambda_i}.$$

We have

$$\sum_{k \le a} \frac{\lambda_i^k}{k!} e^{-\lambda_i} \to \sum_{k \le a} \frac{\lambda^k}{k!} e^{-\lambda} \quad \text{as} \quad i \to +\infty.$$

But

$$\sum_{k \le a} \frac{\lambda^k}{k!} e^{-\lambda} = \sum_{k \le a} \mathbf{P}(X = k) = \mathbf{P}(X \le a).$$

Hence

$$\mathbf{P}(X_i \le a) \to \mathbf{P}(X \le a) \quad \text{as} \quad i \to +\infty.$$

Example 10.14 Let $\lambda > 0$ be given. Let $\{\lambda_i\}$ be a sequence such that $\lambda_i \to \lambda$, $\lambda_i > 0$ and X_i be exponential random variables with parameter λ_i, and let X_i be exponential random variables with parameter λ. Then

$$X_i \to X \quad \text{as} \quad i \to +\infty \quad \text{in distribution.}$$

Theorem 10.15 *Let $\{X_i\}$ be a sequence of random variables with the corresponding m.g.f. $M_i(t)$, and let X be a random variable with the m.g.f. $M(t)$. Assume that there exists $\varepsilon > 0$ such that all these m.g.f. are defined for $t \in (-\varepsilon, \varepsilon)$ and that*

$$M_i(t) \to M(t) \quad as \quad i \to +\infty$$

for all $t \in (-\varepsilon, \varepsilon)$. Then

$$X_i \to X \quad as \quad i \to +\infty \quad in\ distribution.$$

10.5 The Central Limit Theorem

Let

$$\overline{X}_n = \frac{1}{n}\left(X_1 + ... + X_n\right),$$

where X_i are independent random variables with the same distribution.[3] Let $M(t)$ be the m.g.f. for X_i; we consider that it is defined for $t \in (-\varepsilon, \varepsilon)$, where $\varepsilon > 0$. Assume that for all i,

$$\mathbf{E}X_i = \mu, \quad \operatorname{Var} X_i = \sigma^2$$

We found already that

$$\mathbf{E}\overline{X}_n = \mu, \quad \operatorname{Var} \overline{X}_n = \frac{\sigma^2}{n},$$

i.e., that the standard deviation for $\overline{X}_n$ is

$$\overline{\sigma}_n = \sqrt{\operatorname{Var}\left(\overline{X}_n\right)} = \frac{\sigma}{\sqrt{n}}.$$

The idea is that the distribution of $\overline{X}_n$ can be approximated by the Gaussian distribution of

$$\mu + \overline{\sigma}_n Y = \mu + \frac{\sigma}{\sqrt{n}} Y, \quad Y \sim N(0,1).$$

Alternatively, one may say that we want to approximate the distribution of

$$\frac{\overline{X}_n - \mu}{\overline{\sigma}_n}$$

by the distribution of $Y \sim N(0,1)$.

[3]It is often denoted as *iid* random variables, meaning *independent identically distributed.*

Theorem 10.16 *(The Central Limit Theorem) Under the assumptions formulated above,*

$$\frac{\overline{X}_n - \mu}{\overline{\sigma}_n} = \frac{\overline{X}_n - \mu}{\sigma/\sqrt{n}} \to Y \quad as \quad n \to +\infty \qquad in\ distribution,$$

where $Y \sim N(0,1)$. *In other words,*

$$\mathbf{P}\left(\frac{\overline{X}_n - \mu}{\sigma/\sqrt{n}} \le x\right) \to \mathbf{P}(Y \le x) \quad as \quad n \to +\infty$$

for any given x.

Proof. For simplicity, we consider the case where

$$\mu = 0, \quad \sigma = 1.$$

In this case, we have to prove that

$$\frac{\overline{X}_n}{\overline{\sigma}_n} = \frac{\overline{X}_n}{1/\sqrt{n}} \to Y \quad \text{as} \quad n \to +\infty \quad \text{in distribution,}$$

where $Y \sim N(0,1)$.

We have

$$\frac{\overline{X}_n}{\overline{\sigma}_n} = \frac{\overline{X}_n}{1/\sqrt{n}} = \sqrt{n}\overline{X}_n = \frac{\sqrt{n}}{n}(X_1 + ... + X_n) = \frac{1}{\sqrt{n}}(X_1 + ... + X_n).$$

Let $\overline{M}(t)$ be the m.g.f. for

$$\frac{\overline{X}_n}{\overline{\sigma}_n} = \frac{1}{\sqrt{n}}(X_1 + ... + X_n),$$

and $\tilde{M}(t)$ be the m.g.f. for

$$X_1 + ... + X_n.$$

Hence

$$\tilde{M}(t) = M(t) \times \cdots \times M(t) = (M(t))^n,$$

and

$$\overline{M}(t) = \tilde{M}\left(\frac{t}{\sqrt{n}}\right).$$

Note that

$$M(t) = M(0) + M'(0)t + M''(0)\frac{t^2}{2!} + o(t^2),$$

where $o(t^2)$ is a term such that $o(t^2)/t^2 \to 0$ as $t \to 0$.

However,

$$M(0) = 1, \quad M'(0) = \mathbf{E}(X_1) = 0, \quad M''(0) = \mathbf{E}(X_1^2) = 1.$$

Thus

$$M(t) = 1 + \frac{t^2}{2} + o(t^2),$$

where $o(t^2)$ is again a term such that $o(t^2)/t^2 \to 0$ as $t \to 0$.

It follows that

$$M\left(\frac{t}{\sqrt{n}}\right) = 1 + \frac{t^2}{2n} + o\left(\frac{t^2}{n}\right).$$

Hence

$$\overline{M}(t) = \hat{M}\left(\frac{t}{\sqrt{n}}\right) = \left(M\left(\frac{t}{\sqrt{n}}\right)\right)^n = \left(1 + \frac{t^2}{2n} + o\left(\frac{t^2}{n}\right)\right)^n.$$

Lemma 10.17

$$\left(1 + \frac{\alpha}{n}\right)^n \to e^\alpha \quad as \quad n \to +\infty.$$

By this lemma, applied $\alpha = t^2/2$, we obtain

$$\overline{M}(t) = \left(1 + \frac{t^2}{2n} + o\left(\frac{t^2}{2n}\right)\right)^n \to e^{t^2/2} \quad \text{as} \quad n \to +\infty.$$

This completes the proof.

Corollary 10.18 *Let* $Z_n = n\overline{X}_n = X_1 + ... + X_n$. *Then*

$$\frac{Z_n - \mu n}{\sigma\sqrt{n}} \to Y \quad as \quad i \to +\infty \quad in\ distribution,$$

where $Y \sim N(0,1)$.

Proof. It suffices to observe that

$$\mathbf{P}\left(\frac{Z_n - \mu n}{\sigma\sqrt{n}} \le x\right) = \mathbf{P}\left(\frac{\overline{X}_n - \mu}{\sigma/\sqrt{n}} \le x\right).$$

Let $Z_n \sim Bin(n,p)$, i.e., it has binomial distribution with parameters n and p. Consider

$$Z_n = X_1 + ... + X_n,$$

where X_i are independent Bernoulli random variables

$$X_i = \begin{cases} 1, & \text{with probability } p \\ 0, & \text{with probability } 1 - p. \end{cases}$$

It is known that

$$\mathbf{E}X_i = p, \quad \mathbf{E}(X_i)^2 = p, \quad \operatorname{Var}(X_i) = p(1-p).$$

By the Law of Large Numbers,

$$Z_n/n \to p \quad \text{as} \quad i \to +\infty$$

in mean square and in probability. In our notations,

$$\mu = p, \quad \overline{\sigma}_n = \frac{\sigma}{\sqrt{n}}, \quad \sigma^2 = p(1-p).$$

By Central Limit Theorem (see Corollary 10.18), we have that

$$\frac{Z_n - \mu n}{\sigma\sqrt{n}} \to Y \quad \text{in distribution} \quad \text{as} \quad i \to +\infty$$

where $Y \sim N(0,1)$, $\sigma = \sqrt{p(1-p)}$. Let $p = 1/2$, $n = 100$, i.e., $Z_n \sim Bin(100, 1/2)$. Using the theory to estimate $\mathbf{P}(Z_n \le 60)$, we get

$$\mathbf{P}(Z_n \le 60) = \mathbf{P}\left(\frac{Z_n - \mu n}{\sigma\sqrt{n}} \le \frac{60 - \mu n}{\sigma\sqrt{n}}\right).$$

In additions $\mu n = 50$, $\sigma = p(1-p) = 0.25$, $\sqrt{p(1-p)n} = 5$. It gives

$$\mathbf{P}(Z_n \le 60) = \mathbf{P}\left(\frac{Z - \mu n}{\sigma\sqrt{n}} \le \frac{60 - 50}{5}\right) = \mathbf{P}\left(\frac{Z_n - \mu n}{\sigma\sqrt{n}} \le 2\right).$$

Hence

$$\mathbf{P}(Z_n \le 60) \sim \mathbf{P}(Y \le 2).$$

Using table A2 for the standard normal c.d.f., we obtain

$$\mathbf{P}(Z_n \le 60) \sim \mathbf{P}(Y \le 2) = 0.9772.$$

Here we use the closest matching value.

Let $\{X_k\}$ be a sequence of independent random variables uniformly distributed in $[0,1]$ and

$$\overline{X}_n = n^{-1} \sum_{k=1}^{n} X_k.$$

Describe the limit properties of $\overline{X}_n$ as $n \to +\infty$ and estimate $\mathbf{P}(Z_{25} < 0.51)$

Solution. Since X_i are uniformly distributed in $[0,1]$ with the density

$$f(x) = \mathbb{I}_{[0,1]}(x),$$

we have $\mathbf{E}X_i = \int_0^1 x dx = \frac{1}{2}$, $\mathbf{E}(X_i^2) = \int_0^1 x^2 dx = \frac{1}{3}$, and $\sigma^2 = \operatorname{Var} X_i = 1/12$ (see Week 6).

Also,

$$\mathbf{E}\overline{X}_n = n^{-1}\sum_{i=1}^{n}\mathbf{E}X_i = n^{-1}n\frac{1}{2} = \frac{1}{2},$$

$$\operatorname{Var}\overline{X}_n = n^{-2}\sum_{i=1}^{n}\operatorname{Var}X_i = n^{-2}n\frac{1}{12} = \frac{1}{12n}.$$

The Law of Large Numbers is applicable:

$$\overline{X}_n \to \mu = 1/2$$

as $n \to +\infty$ in probability and in mean square, i.e.,

$$\mathbf{P}(|\overline{X}_n - \frac{1}{2}| > \varepsilon) \to 0$$

for any $\varepsilon > 0$, and

$$\mathbf{E}(|\overline{X}_n - \frac{1}{2}|^2) \to 0$$

as $n \to +\infty$.

By the Central Limit Theorem, we obtain the following limit property:

$$\mathbf{P}\left(\frac{\overline{X}_n - 1/2}{1/\sqrt{12n}} \le x\right) \to \mathbf{P}(Y \le x) \quad \text{as} \quad n \to +\infty,$$

where $Y \sim N(0,1)$.

Therefore,

$$\mathbf{P}(Z_n < 0.51) = \mathbf{P}\left(\frac{Z_n - 1/2}{1/\sqrt{12n}} < \frac{0.51 - 1/2}{1/\sqrt{12n}}\right).$$

For $n = 25$, the Central Limit Theorem gives

$$\mathbf{P}(Z_{25} < 0.51) \sim \mathbf{P}\left(Y < \frac{0.01}{1/(5\sqrt{12})}\right).$$

Using table A2 for the standard normal c.d.f., we obtain

$$\mathbf{P}(Z_{25} < 0.51) \sim \mathbf{P}\left(Y < \frac{0.01}{1/(5\sqrt{12})}\right) = \mathbf{P}(Y < 0.1732) = 0.5675.$$

Let $\{X_k\}$ be a sequence of independent random variables uniformly distributed in $[0,1]$. Let

$$Y_k = \min(X_{2k-1}, X_{2k})$$

and

$$\overline{X}_n = n^{-1}\sum_{k=1}^{n} Y_k.$$

Describe the limit properties of $\overline{X}_n$ as $n \to +\infty$ and estimate $\mathbf{P}\left(Z_{25} < \frac{7}{20}\right)$.

Solution. We have X_k are uniformly distributed in $[0,1]$ with the c.d.f.

$$F(x) = x, \qquad x \in [0,1],$$

and the density

$$f(x) = \mathbb{I}_{[0,1]}(x).$$

Hence the density of Y_k is (Week 4)

$$f_Y(y) = 2(1-y)\mathbb{I}_{[0,1]}(y).$$

We obtain

$$\mathbf{E}Y_k = \int_0^1 x2(1-x)dx = 1 - \frac{2}{3} = \frac{1}{3},$$

$$\mathbf{E}\overline{X}_n = n^{-1}\sum_{i=1}^{n} \mathbf{E}Y_k = n^{-1}n\frac{1}{3} = \frac{1}{3}.$$

Further,

$$\mathbf{E}(Y_k^2) = \int_0^1 x^2 2(1-x)dx = 2/3 - \frac{2}{4} = \frac{1}{6},$$

$$\sigma^2 = \operatorname{Var} X_i = 1/6 - 1/9 = 1/18.$$

Let us describe the limit properties.

By the Law of Large Numbers, $\overline{X}_n \to 1/3$ as $n \to +\infty$ in probability, i.e., $\mathbf{P}(|\overline{X}_n - 1/3| > \varepsilon) \to 0$ for any $\varepsilon > 0$.

Further, by the Central Limit Theorem,

$$\mathbf{P}\left(\frac{\overline{X}_n - 1/3}{1/\sqrt{18n}} \leq x\right) \to \mathbf{P}(Y \leq x) \quad \text{as} \quad n \to +\infty,$$

where $Y \sim N(0,1)$.

Now, let us estimate $\mathbf{P}\left(Z_{25} < \frac{7}{20}\right)$. Clearly,

$$\mathbf{P}\left(Z_n < \frac{7}{20}\right) = \mathbf{P}\left(\frac{Z_n - 1/3}{1/\sqrt{18n}} < \frac{7/20 - 1/3}{1/\sqrt{18n}}\right).$$

For $n = 25$, the Central Limit Theorem gives

$$\mathbf{P}(Z_{25} < 7/20) \sim \mathbf{P}\left(Y < \frac{1/60}{1/(5\sqrt{18})}\right).$$

Using table A2 for the standard normal c.d.f., we obtain

$$\mathbf{P}\left(Z_{25} < 7/20\right) \sim \mathbf{P}\left(Y < \frac{1/60}{1/(5\sqrt{18})}\right) = \mathbf{P}\left(Y < 0.3536\right) = 0.6368.$$

(We use the closest matching value.)

Let $\{X_k\}$ be a sequence of independent random variables $N(0,1)$. Let

$$Y_k = X_k^2$$

and

$$S_n = \sum_{k=1}^{n} Y_k, \quad \overline{X}_n = \frac{1}{n}\sum_{k=1}^{n} Y_k.$$

Note that S_n has χ^2 distribution with n degrees of freedom.

Describe the limit properties of $\overline{X}_n$ as $n \to +\infty$ and estimate

$$\mathbf{P}\left(S_{25} < 50\right) = \mathbf{P}\left(Z_{25} < \frac{50}{25}\right) = \mathbf{P}\left(Z_{25} < 2\right).$$

Solution. We know that

$$\mathbf{E}Y_k = 1,$$

$$\mathbf{E}\overline{X}_n = \frac{1}{n}\sum_{i=1}^{n}\mathbf{E}Y_k = \frac{1}{n}n = 1.$$

Further, we found that (Week 9)

$$\mathbf{E}(Y_k^2) = \mathbf{E}(X_k^4) = 3.$$

Hence

$$\sigma^2 = \mathbf{E}(Y_k^2) - (\mathbf{E}Y)^2 = 3 - 1 = 2.$$

To describe the limit properties, we apply the Law of Large Numbers

$$\overline{X}_n \to 1 \quad \text{as} \quad n \to +\infty$$

in probability, i.e., $\mathbf{P}(|\overline{X}_n - 1| > \varepsilon) \to 0$ for any $\varepsilon > 0$.

Further, by the Central Limit Theorem,

$$\mathbf{P}\left(\frac{\overline{X}_n - 1}{\sqrt{2}/\sqrt{n}} \le x\right) \to \mathbf{P}(Y \le x) \quad \text{as} \quad n \to +\infty,$$

where $Y \sim N(0,1)$. Clearly,

$$\mathbf{P}\left(Z_n < 2\right) = \mathbf{P}\left(\frac{Z_n - 1}{\sqrt{2}/\sqrt{n}} < \frac{2-1}{\sqrt{2}/\sqrt{n}}\right).$$

For $n = 25$, it gives

$$\mathbf{P}\left(Z_{25} < 2\right) \sim \mathbf{P}\left(Y < \frac{1}{\sqrt{2}/5}\right) = \mathbf{P}\left(Y < \frac{5}{\sqrt{2}}\right).$$

Using table A2 for the standard normal c.d.f., we obtain

$$\mathbf{P}\left(S_{25} < 50\right) = \mathbf{P}\left(Z_{25} < 2\right) \sim \mathbf{P}\left(Y < 5/\sqrt{2}\right) = \mathbf{P}\left(Y < 3.53\right) = 0.9989.$$

Problems for Week 10

Problem 10.1 *Let $\{X_k\}$ be a sequence of independent random variables with the density*

$$f_X(x) = 2x\mathbb{I}_{[0,1]}(x).$$

Let

$$Z_n = n^{-1}\sum_{k=1}^{n} X_k.$$

(a) Find $\mathbf{E}(Z_n)$ and $\mathrm{Var}(Z_n)$ for all n.
(b) Describe the limit properties of Z_n as $n \to +\infty$. (Hint: use appropriate limit theorems.)
(c) Estimate $\mathbf{P}(Z_{16} > 0.7)$.

Problem 10.2 *Let $\{X_k\}$ be a sequence of independent random variables with the density*

$$f(x) = 5e^{-5x}\mathbb{I}_{[0,+\infty]}.$$

Let

$$Z_n = n^{-1}\sum_{k=1}^{n} X_k.$$

(a) Find $\mathbf{E}(Z_n)$ and $\mathrm{Var}(Z_n)$ for all n.
(b) Describe the limit properties of Z_n as $n \to +\infty$. (Hint: use appropriate limit theorems.)
(c) Estimate $\mathbf{P}(Z_{11} > 0.25)$.

Problem 10.3 *Consider a casino roulette game. In this game, players make bets on certain numbers. The probability that the number 1 come up in one game is 1/38.*

(a) Assume that one bets \$1 on a single number, say, number 1. If this number comes up, then the gain is \$35. On the other hand, if this number does not come up, then \$1 will be the loss, i.e., the gain is −\$1. Find the expectation and the variance for the gain.
(b) Assume that one makes 20 games described in (a). Using Central Limit Theorem as an approximation, estimate the probability that the average gain is positive.

Problem 10.4 *Consider a casino roulette game. In this game, players make bets on the red. The probability that the red comes up in one game is* 18/38.

(a) Assume that one bets \$1 *on the red. If this color comes up, then the gain is* \$1. *On the other hand, if this color does not come up, then* \$1 *will be the loss, i.e. the gain is* −\$1. *Find the expectation and the variance for the gain.*

(b) Assume that one makes 20 *games described in (a). Using Central Limit Theorem as an approximation, estimate the probability that the average gain is positive.*

Week 11. Statistical Inference: Point Estimation

Consider a sample $(X_1, ..., X_n)$. Assume that this sample is obtained as n measurements of a random variable X in some experiment, therefore X_i are random variables themselves, with the same distribution as X. We assume the experiment is such that X_i are mutually independent and have the same distribution, i.e., an i.i.d. sample (independent and identically distributed sample). We call the distribution of X the population distribution. Respectively, we call the variance of X the population variance, the moments of X the population moments, etc.

In practice, the population distribution is often unknown. Assume that there is a hypothesis about the population distribution of X, i.e., X has a probability density function or a probability frequency function $f(x|\theta)$ that depends on some unknown parameter θ (it can be a vector).

Estimation of θ from observations X_i is called statistical inference. This topic will be discussed below.

11.1 Maximum Likelihood Estimation

An estimate of the value of the unknown parameter that fits the distribution can be obtained via so-called point estimation. This can be achieved via the methods of moments and the Maximum Likelihood Estimation (MLE). In this section, we consider the MLE method.

Consider a sample $X = (X_1, ..., X_n)$. Assume that this sample is obtained as n measurements of a random variable X. Let X have the density $f(x|\theta)$ that depends on unknown parameter θ.

The maximum likelihood function is

$$L(\theta) = \prod_{i=1}^{n} f(X_i|\theta).$$

The idea is to take the solution of the problem of maximization $L(\theta)$ as the estimate $\hat{\theta}$ of θ. This is the Maximum Likelihood Estimation (MLE) method.

The following rule can be useful: if $U'(x) = 0$ and $U''(x) < 0$ then x is the point of (local) maximum of U.

The MLE method usually works well for the case where $f(x|\theta)$ has the only maximum in θ given x. The method requires to take the following steps:

(a) Write down the likelihood function for the available data.
(b) Write down the log-likelihood function

$$\ell(\theta) = \log L(\theta) = \sum_{i=1}^{n} \log f(X_i|\theta).$$

This will usually simplify the algebra.
(c) Find a point of maximum for the log-likelihood function.
(d) Check that the values you have found do maximize the likelihood function. (This will usually involve calculation of the second derivative.)

Theorem 11.1 *(Invariance property of MLE) If θ is an MLE of θ and $g(\cdot)$ is a function, then $g(\hat{\theta})$ is an MLE of $g(\theta)$.*

Example: MLE for normal distribution with known σ^2. Assume the hypothesis that $X \sim N(\mu, \sigma^2)$, with some known σ^2 and unknown μ. Let us estimate μ using the MLE method. We have

$$L(\mu) = \prod_{i=1}^{n} f(X_i|\mu) = \left[\frac{1}{\sqrt{2\pi\sigma^2}}\right]^n \exp\left(-\sum_{i=1}^{n} \frac{(X_i - \mu)^2}{2\sigma^2}\right)$$

and

$$\ell(\mu) = \log L(\mu) = \log \prod_{i=1}^{n} f(X_i|\mu) = n \log\left[\frac{1}{\sqrt{2\pi\sigma^2}}\right] - \sum_{i=1}^{n} \frac{(X_i - \mu)^2}{2\sigma^2}.$$

Thus,

$$\frac{d\ell(\mu)}{d\mu} = \frac{1}{2\sigma^2}\sum_{i=1}^{n} 2(X_i - \mu) = \frac{1}{\sigma^2}\left(\sum_{i=1}^{n} X_i - n\mu\right).$$

The equation

$$\frac{d\ell(\mu)}{d\mu} = 0$$

gives solution

$$\hat{\mu} = \frac{1}{n}\sum_{i=1}^{n} X_i.$$

Further, we have

$$\frac{d^2\ell(\mu)}{d\mu^2} = -\frac{1}{\sigma^2}n < 0.$$

Hence $\hat{\mu}$ is the MLE. Assume that $n = 3$, $X_1 = 0.7$, $X_2 = 0.2$, $X_3 = 0.3$. Then

$$\hat{\mu} = \frac{1}{3}(0.7 + 0.2 + 0.3) = 0.4.$$

Example: MLE for normal distribution with unknown μ, σ^2. Assume the hypothesis that $X \sim N(\mu, \sigma^2)$, with unknown (μ, σ^2). MLE estimate of (μ, σ^2) is

$$\hat{\mu} = \overline{x} = \frac{1}{n}\sum_{i=1}^{n} X_i, \quad \hat{\sigma}^2 = \frac{1}{n}\sum_{i=1}^{n}(X_i - \overline{x})^2.$$

Note that it is different from the sample variance defined above as

$$S^2 = \frac{1}{n-1}\sum_{i=1}^{n}(X_i - \overline{x})^2.$$

S^2 is the so-called unbiased estimate; it is close to $\hat{\sigma}^2$ for large n.

Example: MLE for exponential distribution. Assume the hypothesis that $X \sim Exp(\lambda)$, with unknown λ. Let us estimate λ using MLE. We have

$$L(\lambda) = \prod_{i=1}^{n} f(X_i|\lambda) = \lambda^n \prod_{i=1}^{n} \exp(-\lambda X_i)$$

and

$$\ell(\lambda) = \log L(\lambda) = \log \prod_{i=1}^{n} f(X_i|\lambda) = n\log\lambda - \sum_{i=1}^{n} \lambda X_i.$$

Thus

$$\frac{d\ell(\lambda)}{d\lambda} = \frac{n}{\lambda} - \sum_{i=1}^{n} X_i = \frac{n}{\lambda} - n\overline{X},$$

where $\overline{X} = \frac{1}{n}\sum_{i=1}^{n} X_i$ is the sample mean.

The equation

$$\frac{d\ell(\lambda)}{d\lambda} = 0$$

gives solution $\hat{\lambda} = 1/\overline{x}$. Further, we have

$$\frac{d^2\ell(\lambda)}{d\lambda^2} = -\frac{n}{\lambda^2} < 0.$$

Hence $\hat{\lambda}$ is the MLE. Assume that $n = 3$, $X_1 = 0.7$, $X_2 = 0.2$, $X_3 = 0.3$. Then

$$\overline{X} = \frac{1}{3}(0.7 + 0.2 + 0.3) = 0.4, \quad \hat{\lambda} = 1/0.4 = 2.5.$$

Note that

$$\mathbf{E}(X|\hat{\lambda} = 2.5) = 1/2.5 = 0.4 = \overline{X}.$$

11.2 The method of moments

In this section we discuss the method of moments.

Consider a sample $X_1, ..., X_n$ obtained as n measurements of a random variable X. Assume that the distribution of X depends on a parameter θ.

Definition 11.2 *Let $k = 1, 2, ...$ The* sample moments *are calculated as*

$$m_k = \frac{1}{n}\sum_{i=1}^{n} X_i^k.$$

The value $\overline{X} = \mu_1 = \overline{X}_n = \frac{1}{n}(X_1 + ... + X_n)$ is said to be the sample mean.

We know that, by the Law of Large Numbers,

$$\mu_k = \frac{1}{n}(X_1^k + ... + X_n^k) \to \mathbf{E}X_i^k \quad \text{as} \quad n \to +\infty$$

in probability and in mean square. Therefore, $\mathbf{E}X_i^k$ can be estimated as μ_k for large n.

Assume that a hypothesis about the family of the distributions is accepted, i.e., that X has a probability density function or probability frequency function $f(x|\theta)$ that depends on some unknown parameter θ (it can be a vector). In this case, the population moments

$$\mu'_k = \mu'_k(\theta) = \mathbf{E}(X^k|\theta)$$

are function of θ.

We equate the population moments and the sample moments, then solve these equations to obtain unknown parameters of the distribution. In other words, we estimate θ by solving the system

$$\mu'_k(\theta) = m_k, \quad k = 1, ..., d,$$

where d is the dimension of θ. We use as many equations as required.

In this chapter, we will be using sample mean and sample variance as estimates for $\mathbf{E}X$ and $\operatorname{Var} X$, respectively.

Example: method of moments for uniform distribution. Assume that X has uniform distribution on $[0, \theta]$. Assume that $X_1 = 0.7$, $X_2 = 0.2$, $X_3 = 0.3$. Let us estimate θ using the method of moments. We have

$$m_1 = \frac{1}{3}(0.7 + 0.2 + 0.3) = 0.4$$

and

$$\mu'_1(\theta) = \mathbf{E}(X|\theta) = \frac{1}{\theta}\int_0^\theta x dx = \frac{\theta}{2}.$$

It gives one equation

$$\frac{\theta}{2} = 0.4.$$

Thus, we obtain the estimate $\hat{\theta} = 0.8$.

Example: method of moments for normal distribution. Assume the hypothesis that $X \sim N(\mu, \sigma^2)$, with unknown $\theta = (\mu, \sigma^2)$ and $X_1 = 0.7$, $X_2 = 0.2$, $X_3 = 0.3$. Let us estimate θ using the method of moments. We have

$$m_1 = \frac{1}{3}(0.7 + 0.2 + 0.3) = 0.4,$$

$$m_2 = \frac{1}{3}(0.7^2 + 0.2^2 + 0.3^2) = 0.2067$$

and

$$\mu'_1(\theta) = \mathbf{E}(X|\theta) = \mu, \quad \mu'_2(\theta) = \mathbf{E}(X^2|\theta) = \sigma^2 + \mu^2.$$

It gives one equation

$$\hat{\mu} = 0.4, \quad \hat{\sigma}^2 = 0.2067 - 0.4^2 = 0.0467.$$

Thus, we obtain the estimate $\hat{\theta} = (\hat{\mu}, \hat{\sigma}^2) = (0.4, 0.0467)$.

There are modifications of the method of moments that use other quantitative characteristics of the distributions instead of moments.

For example, the method of quantiles (or method of percentiles) is a modification of method of moments such that, instead of moments, we use the quantiles, i.e., we equate population and sample quantiles.

11.3 Sample variance

Let us consider estimation of σ^2. By the definition,

$$\sigma^2 = \mathrm{Var}\,(X_i) = \mathbf{E}(X_i - \mu)^2.$$

Assume that μ is known. In this case, we can use the Law of Large Numbers again:

$$\frac{1}{n}((X_1-\mu)^2 + ... + (X_n-\mu)^2) \to \sigma^2 \quad \text{as} \quad n \to +\infty$$

in probability and in mean square.

Unfortunately, we cannot use this if μ is unknown and has to be estimated.

Definition 11.3 *The value*

$$S^2 = \frac{1}{n-1}\left((X_1 - \overline{X}_n)^2 + ... + (X_n - \overline{X}_n)^2\right)$$

is said to be the sample variance.

The sample variance is a random variable. It is commonly used as an approximation of σ^2.

Let $n = 3$ and $(X_1, X_2, X_3) = (-1, 0, 2)$. Then

$$\overline{X} = \frac{1}{3}(-1+0+2) = \frac{1}{3},$$

$$S^2 = \frac{1}{3-1}\left(\left(-1-\frac{1}{3}\right)^2 + \left(0-\frac{1}{3}\right)^2 + \left(2-\frac{1}{3}\right)^2\right) = 2.333.$$

11.4 Properties of the sample mean and sample variance

Lemma 11.4 *The correlation of $\overline{X}$ with any of the random variables $(X_1 - \overline{X})$, $(X_2 - \overline{X})$,..., $(X_n - \overline{X})$ is zero.*

Proof. We consider only the case where $n = 2$. It suffices to prove that

$$\mathrm{Cov}(\overline{X}, X_i - \overline{X}) = 0.$$

We have

$$\overline{X} = \frac{1}{2}(X_1 + X_2).$$

Hence

$$X_1 - \overline{X} = \frac{1}{2}(X_1 - X_2), \qquad X_2 - \overline{X} = \frac{1}{2}(X_2 - X_1)$$

and

$$\mathrm{Cov}(\overline{X}, X_1 - \overline{X}) = \mathrm{Cov}\left(\frac{1}{2}(X_1 + X_2), \frac{1}{2}(X_1 - X_2)\right).$$

Also,

$$\mathbf{E}\frac{1}{2}(X_1 - X_2) = 0, \quad \mathbf{E}\frac{1}{2}(X_2 - X_1) = 0.$$

Therefore,

$$\mathrm{Cov}(\overline{X}, X_1 - \overline{X}) = \mathbf{E}\left(\frac{1}{2}(X_1 + X_2), \frac{1}{2}(X_1 - X_2)\right)$$
$$= \frac{1}{4}\mathbf{E}\Big((X_1 + X_2)(X_1 - X_2)\Big)$$
$$= \frac{1}{4}\Big(\mathbf{E}(X_1X_2) + \mathbf{E}(X_1^2) - \mathbf{E}(X_1X_2) - \mathbf{E}(X_2^2)\Big) = 0.$$

Similarly, it can be shown that

$$\mathrm{Cov}(\overline{X}, X_2 - \overline{X}) = 0.$$

11.5 Properties of Gaussian samples

Starting from now and up to the end of these lectures, we assume that X_i are Gaussian,

$$X_i \sim N(\mu, \sigma^2).$$

Theorem 11.5 $\overline{X} \sim N\left(\mu, \frac{\sigma^2}{n}\right)$.

Proof. We know that $\mathbf{E}\overline{X} = \mu$ and $\mathrm{Var}\,\overline{X} = \sigma^2/n$. In addition, we know that $\overline{X}$ is a sum of Gaussian random variables therefore is Gaussian.

Theorem 11.6 *If X_i are Gaussian then $\overline{X}$ is independent from the random variables $(X_1 - \overline{X})$, $(X_2 - \overline{X})$,..., $(X_n - \overline{X})$.*

Proof. We already proved that the correlation of $\overline{X}$ with the random variables $(X_1 - \overline{X})$,$(X_2 - \overline{X})$,..., $(X_n - \overline{X})$ is zero. Then the independence follows from the property of Gaussian random variables.

Theorem 11.7 *$\overline{X}$ and S^2 are independent.*

Proof. We have that S^2 is a function of random variables $(X_1-\overline{X})$, $(X_2-\overline{X})$,..., $(X_n - \overline{X})$ that are independent from $\overline{X}$. Hence S^2 is independent from $\overline{X}$.

Theorem 11.8

$$\frac{n-1}{\sigma^2}S^2 \sim \chi^2_{n-1},$$

i.e., this random variable has χ^2_{n-1} distribution.

Proof. Let us assume for simplicity that $\sigma = 1$. We have that

$$\sum_{i=1}^{n}(X_i - \mu)^2 = \sum_{i=1}^{n}(X_i - \overline{X} + \overline{X} - \mu)^2.$$

Expanding, we obtain

$$\begin{aligned}\sum_{i=1}^{n}(X_i - \mu)^2 &= \sum_{i=1}^{n}\left[(X_i - \overline{X})^2 + 2(X_i - \overline{X})(\overline{X} - \mu) + (\overline{X} - \mu)^2\right] \\ &= \sum_{i=1}^{n}(X_i - \overline{X})^2 + 2\sum_{i=1}^{n}(X_i - \overline{X})(\overline{X} - \mu) + \sum_{i=1}^{n}(\overline{X} - \mu)^2.\end{aligned}$$

On the other hand,

$$\begin{aligned}2\sum_{i=1}^{n}(X_i - \overline{X})(\overline{X} - \mu) &= 2(\overline{X} - \mu)\sum_{i=1}^{n}(X_i - \overline{X}) \\ &= 2(\overline{X} - \mu)\left(\sum_{i=1}^{n}X_i - n\overline{X}\right) = 0,\end{aligned}$$

since

$$\sum_{i=1}^{n}X_i - n\overline{X} = 0.$$

Therefore

$$\begin{aligned}\sum_{i=1}^{n}(X_i - \mu)^2 &= \sum_{i=1}^{n}(X_i - \overline{X})^2 + \sum_{i=1}^{n}(\overline{X} - \mu)^2 \\ &= (n-1)S^2 + n(\overline{X} - \mu)^2 \\ &= (n-1)S^2 + (\sqrt{n}(\overline{X} - \mu))^2.\end{aligned}$$

Note that, for $\sigma = 1$,

$$X_i - \mu \sim N(0,1), \quad \overline{X} \sim N\left(\mu, \frac{1}{n}\right).$$

Hence

$$\overline{X} - \mu \sim N\left(0, \frac{1}{n}\right), \quad \sqrt{n}(\overline{X} - \mu) \sim N(0,1).$$

This gives

$$U = (\sqrt{n}(\overline{X} - \mu))^2 \sim \chi_1^2.$$

Therefore

$$\sum_{i=1}^{n}(X_i - \mu)^2 \sim \chi_n^2.$$

Also, we know that $U = (\sqrt{n}(\overline{X} - \mu))^2$ is independent from S^2, and $(n-1)S^2 + U \sim \chi_n^2$. It follows that

$$(n-1)S^2 = \frac{n-1}{\sigma^2}S^2 \sim \chi_{n-1}^2.$$

Example 11.9 Let $n = 5$, $X_i \sim N(2,3)$. Then the distribution of $\overline{X}$ is $N(2, \frac{3}{5})$ and the distribution of $\frac{5-1}{3}S^2 = \frac{4}{3}S^2$ is χ_4^2. Find $\mathbf{P}\left(S^2 < 1\right)$.

Solution. We have

$$\mathbf{P}\left(S^2 < 1\right) = \mathbf{P}\left(\frac{4}{3}S^2 < \frac{4}{3}\right) = \mathbf{P}\left(U < 1.33\right) \sim 0.1,$$

where $U \sim \chi_4^2$ (with degree of freedom 4). We use table A3 for χ_4^2. In the first row, this table shows p such that $p = \mathbf{P}(U > c)$, where $U \sim \chi_v^2$ (i.e., with v degrees of freedom).

degrees of freedom (v)	..	$p = 0.9$	....	....
..	...	...	....	...
4	...	1.064	...	...

Obviously, we can find

$$\mathbf{P}(U \leq c) = \mathbf{P}(U < c) = 1 - \mathbf{P}(U < c) = 1 - p.$$

Hence

$$\begin{aligned}\mathbf{P}\left(S^2 < 1\right) &= \mathbf{P}\left(U < 1.33\right) \\ &= 1 - \mathbf{P}\left(U > 1.33\right) \sim 1 - \mathbf{P}(U > 1.026) \\ &= 1 - 0.9 = 0.1.\end{aligned}$$

Example 11.10 Let $n = 5$, $X_i \sim N(0,9)$. Then the distribution of $\overline{X}$ is $N(0, \frac{9}{5})$ and the distribution of $\frac{5-1}{9}S^2 = \frac{4}{9}S^2$ is χ_4^2. Find $\mathbf{P}\left(S^2 < 1\right)$.

Solution. We have

$$\mathbf{P}\left(S^2 < 1\right) = \mathbf{P}\left(\frac{4}{9}S^2 < \frac{4}{9}\right) = \mathbf{P}\left(U < 0.444\right),$$

where $U \sim \chi_4^2$ (with four degrees of freedom). We use table A3 for χ_4^2.

degrees of freedom (v)	..	$p = 0.975$	....	....
..	...	...	....	...
4	...	0.484	...	...

This gives

$$\begin{aligned}\mathbf{P}\left(S^2 < 1\right) &= \mathbf{P}\left(U < 0.444\right)\\ &= 1 - \mathbf{P}(U > 0.444) \sim 1 - \mathbf{P}(U > 0.484) = 0.025.\end{aligned}$$

Example 11.11 *Let $n = 4$, $X_i \sim N(0, 0.75)$. In this case, $\frac{4-1}{0.75}S^2 = 4S^2 \sim \chi_3^2$. Let us find $\mathbf{P}(S^2 < 2)$.*

Solution. We have that

$$\mathbf{P}(S^2 < 2) = \mathbf{P}(4S^2 < 8) \sim \mathbf{P}(U < 8),$$

where $U \sim \chi_3^2$ (with three degrees of freedom).

We use table A3 for χ_3^2.

degrees of freedom (v)	..	$p = 0.05$	....	....
..	...	...	....	...
3	...	7.815	...	...

We can find

$$\mathbf{P}(U \le c) = \mathbf{P}(U < c) = 1 - \mathbf{P}(U > c) = 1 - p.$$

Hence

$$\begin{aligned}\mathbf{P}\left(S^2 < 1\right) = \mathbf{P}\left(U < 8\right) = 1 - \mathbf{P}(U > 8) \approx 1 - \mathbf{P}(U > 7.82) &\sim 1 - 0.05\\ &= 0.95.\end{aligned}$$

Theorem 11.12 $\mathbf{E}S^2 = \sigma^2$.

Proof. For χ_{n-1}^2 distribution, the expectation is $n - 1$. It follows that

$$\mathbf{E}\left(\frac{n-1}{\sigma^2}S^2\right) = \frac{n-1}{\sigma^2}\mathbf{E}\left(S^2\right) = n - 1.$$

It gives $\mathbf{E}S^2 = \sigma^2$. This is why S^2 is said to be an unbiased estimate for σ^2.

11.6 Some useful distributions for Gaussian samples

Student's distribution (t distribution)

Definition 11.13 *Let*

$$Z \sim N(0,1), \quad U \sim \chi_n^2,$$

and Z and U are independent. The distribution of

$$\frac{Z}{\sqrt{U/n}}$$

is said to be Student's t-distribution with n degrees of freedom (or simply t-distribution with n degrees of freedom), and it is denoted as t_n.

Let $\overline{X}$ and S^2 be defined as above for independent random variables $X_1, ..., X_n$, $X_i \sim N(\mu, \sigma^2)$.

Theorem 11.14 *Student's distribution with n degrees of freedom has the density*

$$f(t) = c\left(1 + t^2/n\right)^{-(n+1)/2}, \quad t \in (-\infty, +\infty),$$

where $c > 0$ is a constant.

In fact, we do not need to specify c in the above equation for the density; it is uniquely defined from the property

$$\int_{-\infty}^{\infty} f(t)dt = 1.$$

The equation for f can be rewritten as

$$f(t) \propto \left(1 + t^2/n\right)^{-(n+1)/2}, \quad t \in (-\infty, +\infty).$$

The last expression uses the proportionality sign "$\propto$" to indicate that the density f is proportional to the given function, i.e., that these functions are the same up to multiplication by a constant. This popular and convenient notation is commonly used in Probability and Statistics.

The t_n distribution is symmetric about zero. As $n \to +\infty$, this distribution tends to the standard normal distribution; in fact, for $n \geq 20$, the distributions are very close.

Theorem 11.15 *The random variable*

$$\frac{\overline{X} - \mu}{\sqrt{S^2/n}}$$

has t_{n-1}-distribution, i.e., Student's distribution with $n-1$ degrees of freedom.

Proof. Observe that

$$\frac{S^2}{\sigma^2} = \frac{(n-1)S^2}{\sigma^2(n-1)} = \frac{U}{n-1},$$

where $U \sim \chi^2_{n-1}$. Hence

$$\frac{\overline{X}-\mu}{\sqrt{S^2/n}} = \frac{\sqrt{n}\,\frac{\overline{X}-\mu}{\sigma}}{\sqrt{n}\sqrt{S^2/n\sigma^2}} = \frac{\frac{\overline{X}-\mu}{\sigma/\sqrt{n}}}{\sqrt{S^2/\sigma^2}} = \frac{\frac{\overline{X}-\mu}{\sigma\sqrt{n}}}{\sqrt{U/(n-1)}}.$$

We know that

$$\frac{\overline{X}-\mu}{\sigma\sqrt{n}} \sim N(0,1).$$

Then the proof follows.

Note that that the expectation for these distributions is not defined and that the number of finite moments of t_n depends on n.

Cauchy density

Suppose that X and Y are independent standard normal random variables $N(0,1)$ and $Z = Y/X$. We found in Week 4 that

$$f_z(z) = \frac{1}{\pi(1+z^2)}.$$

This density is called the Cauchy density. Cauchy distribution is t_1 distribution. Note that expectation for this distribution is not defined.

F-distribution

Definition 11.16 *Let U and V be independent chi-square random variables with m and n degrees of freedom, respectively. The distribution of*

$$W = \frac{U/m}{V/n}$$

is called the F distribution with m and n degrees of freedom and is denoted by $F_{m,n}$.

Theorem 11.17 *$F_{m,n}$-distribution has the density*

$$f(w) = cw^{m/2-1}\left(1+\frac{m}{n}w\right)^{-(m+n)/2}, \quad w \geq 0$$

where $c > 0$ is a constant.

In other words,

$$f(w) \propto w^{m/2-1}\left(1+\frac{m}{n}w\right)^{-(m+n)/2}, \quad w \geq 0.$$

It can be shown that, for $n > 2$, $\mathbf{E}(W)$ exists and is finite and $\mathbf{E}(W) = n/(n-2)$.

Theorem 11.18 *Let $T \sim t_n$. Then $T^2 \sim F_{1,n}$. Let $X \sim F_{m,n}$. Then $X^{-1} \sim F_{n,m}$. If X and Y are independent exponential random variables $Exp(1)$, then $Z = X/Y$ follows an $F_{2,2}$ distribution.*

Proof.

$$f_z(z) = \int_{-\infty}^{\infty} |x| f_X(x) f_Y(zx) dx = \int_0^{\infty} x e^{-x} e^{-zx} dx$$
$$= \int_0^{\infty} x e^{-(z+1)x} dx = \frac{1}{1+z}\int_0^{\infty} x(1+z)e^{-(z+1)x} dx.$$

We have that

$$\int_0^{\infty} t^{\alpha} \lambda e^{-\lambda t} dt = \frac{\Gamma(\alpha+1)}{\lambda^{\alpha}}.$$

Hence

$$f_z(z) = \frac{1}{1+z}\frac{\Gamma(2)}{(1+z)} = \frac{1}{(1+z)^2}.$$

It is a $F_{m,n}$-distribution with $m/2 - 1 = 0$, $-(m+n)/2 = -2$. It gives $m = n = 2$.

***F* distribution for the ratios of the sample variances**

Let U and V be independent chi-square random variables with n and m degrees of freedom, respectively. The distribution of

$$W = \frac{U/n}{V/m}$$

is called the F distribution with m and n degrees of freedom and is denoted by $F_{n,m}$.

Theorem 11.19 *Consider two i.i.d. samples, $X_1, ..., X_n$, and $Y_1, ..., Y_m$. Assume that these samples are independent where $X_i \sim N(\mu, \sigma_X^2)$ and $Y_i \sim N(\mu, \sigma_Y^2)$. Let S_X^2 and S_Y^2 be the corresponding sample variances. Then*

$$\frac{\sigma_Y^2 S_X^2}{\sigma_X^2 S_Y^2} \sim F_{n-1,m-1}.$$

The proof of this theorem follows immediately from the definitions of the corresponding distributions.

Problems for Week 11

Problem 11.1 *Consider independent identically distributed random variables X_1, X_2, X_3. Let $\overline{X}$ be the sample mean, and let S^2 be the sample variance.*

(a) Assume that, in an experiment, we obtain values $(X_1, X_2, X_3) = (-1, -1, 1)$. Calculate the corresponding values $\overline{X}$ and S^2.
(b) Assume that X_i belong to the normal distribution $N(2, 3)$. Describe the distribution of $\overline{X}$ and estimate $\mathbf{P}(\overline{X} < 2.2)$ for this case.
(c) Assume that X_i belong to the normal distribution $N(2, 3)$. Describe the distribution of S^2 and estimate $\mathbf{P}(S^2 < 6.9)$ for this case.

Problem 11.2 *Consider independent identically distributed random variables $X_1, X_2, X_3, X_4, X_5, X_6$. Let $\overline{X}$ be the sample mean, and let S^2 be the sample variance.*

(a) Assume that, in an experiment, we obtain values

$$(X_1, X_2, X_3, X_4, X_5, X_6) = (-1, -1, 0, 2, 1, 2).$$

Calculate the corresponding values $\overline{X}$ and S^2.
(b) Assume that X_i belong to the normal distribution $N(2, 4)$. Describe the distribution of $\overline{X}$ and estimate $\mathbf{P}(\overline{X} < 2.2)$ for this case.
(c) Assume that X_i belong to the normal distribution $N(2, 4)$. Describe the distribution of S^2 and estimate $\mathbf{P}(S^2 < 0.484)$ for this case.

Problem 11.3 *Consider independent identically distributed random variables $X_1, X_2, X_3, X_4, X_5, X_6$. Let $\overline{X}$ be the sample mean, and let S^2 be the sample variance. Assume that X_i belong to the normal distribution $N(2, 3)$.*

(a) Describe the distribution of $\overline{X}$ and estimate $\mathbf{P}(\overline{X} < 2.2)$.
(b) Describe the distribution of S^2 and estimate $\mathbf{P}(S^2 < 6)$.

Week 12. Statistical Inference: Interval Estimation

In this chapter, we review statistical inference and hypothesis testing based on interval estimation.

12.1 Critical values for the distributions

Critical values and quantiles for the normal distribution

For $\alpha \in (0,1)$, the real numbers $z_{\alpha/2}$ are called the critical values for the standard normal distribution if

$$\mathbf{P}(Z > z_{\alpha/2}) = \alpha/2,$$

where $Z \sim N(0,1)$. It follows that

$$\mathbf{P}(-z_{\alpha/2} \le Z \le z_{\alpha/2}) = 1 - \alpha.$$

They are defined from the tables. In additions, $q_{1-\alpha/2} = z_{\alpha/2}$ and $q_{\alpha/2} = -z_{\alpha/2}$, where q_β are the quantiles, i.e. $\mathbf{P}(Z \le q_\beta) = \beta$; in particular,

$$\mathbf{P}(Z \le q_{1-\alpha/2}) = 1 - \alpha/2, \qquad \mathbf{P}(Z \le q_{\alpha/2}) = \alpha/2.$$

For $\alpha = 0.05$ we have $q_{1-\alpha/2} = z_{\alpha/2} = 1.65$, $q_{\alpha/2} = -z_{\alpha/2} = -1.65$.

Critical values for the chi-square and other distributions

For $\alpha \in (0,1)$, the real numbers $\chi^2_{1-\alpha/2}$ and $\chi^2_{\alpha/2}$ are called the critical values for the chi-square distribution if

$$\mathbf{P}(X < \chi^2_{1-\alpha/2}) = \alpha/2, \quad \mathbf{P}(X > \chi^2_{\alpha/2}) = \alpha/2,$$

where $X \sim \chi^2_n$. For these values,

$$\mathbf{P}(\chi^2_{1-\alpha/2} \le X \le \chi^2_{\alpha/2}) = 1 - \alpha.$$

The value $\chi^2_{1-\alpha/2}$ is called the lower-tail critical value and the value $\chi^2_{\alpha/2}$ is called the upper-tail critical value. Note that the choice of these notations is traditional and it could be confusing; in particular, we have that $\chi^2_{0.95} < \chi^2_{0.05}$. The critical values for the t and F-distributions are defined similarly. For instance, if $Y \sim t_n$, then

$$\mathbf{P}(-\theta^2_{n,\alpha/2} \le Y \le t_{\alpha/2}) = 1 - \alpha, \quad \mathbf{P}(Y \le t_{n,\alpha}) = 1 - \alpha),$$

where $t_{n,\alpha}$ is a critical value for the t_n-distribution. In the tables for $F_{n,m}$-distributions, the corresponding critical values $f_\alpha = f^{n,m}_\alpha$ for the distributions are usually given for $\alpha \in (0.5, 1)$ only. $F_{n,m}$ only; this is actually sufficient since

$$f^{n,m}_{1-\alpha} = (f^{m,n}_\alpha)^{-1}.$$

12.2 Interval estimation: Confidence interval

Usually, it is impossible to restore the exact value of θ without an error, even from a large statistical sample. Therefore, the presentation of the results of the statistical inference should accommodate some description of the remaining uncertainty. For this purpose, it is common to use the so-called interval estimate of θ. More precisely, for a sample $X = (X_1, ..., X_n)$, we construct a random interval $[L(X), R(X)]$ depending on the sample which contains θ with certain preselected probability.

If the statistics $L(X)$ and $R(X)$ are such that $\mathbf{P}(\theta \in [L(X), R(X)]) \ge 1 - \alpha$ then the interval $[L(X), R(X)]$ is called a $100(1-\alpha)\%$-confidence interval for θ.

Examples of confidence intervals

Estimation of the mean

Consider an i.i.d. sample $X_1, ..., X_n$ from a population distribution with unknown mean μ. Let us assume that the variance σ^2 is known.

We estimate the mean as the sample mean $\overline{X}$. We have that $\mathbf{E}\overline{X} = \mu$ and $\operatorname{Var}\overline{X} = \sigma^2/n$. Now, let $E = \overline{X} - \mu$ be the estimation error.

Typically, the distribution of E is not presented in the list of the distributions with available tables. In this case, we may use the Central Limit Theorem. By this theorem, we have that

$$\frac{E}{\sigma/\sqrt{n}} = \frac{\overline{X} - \mu}{\sigma/\sqrt{n}}$$

has the distribution that is close to the distribution $N(0,1)$ for large enough n. Hence we accept that

$$\mathbf{P}(-z_{\alpha/2} \leq \frac{E}{\sigma/\sqrt{n}} \leq z_{\alpha/2}) = 1-\alpha,$$

(approximately, for large enough n), or

$$\mathbf{P}(-\frac{z_{\alpha/2}\sigma}{\sqrt{n}} \leq E \leq \frac{z_{\alpha/2}\sigma}{\sqrt{n}}) = 1-\alpha.$$

Therefore, $[-z_{\alpha/2}\sigma/\sqrt{n}, z_{\alpha/2}\sigma/\sqrt{n}]$ is $(1-\alpha)100\%$-confidence interval for the estimation error. Let us construct the $(1-\alpha)100\%$-confidence interval for the estimate. We accept that

$$\mathbf{P}(-z_{\alpha/2} \leq \frac{\overline{X}-\mu}{\sigma/\sqrt{n}} \leq z_{\alpha/2}) = 1-\alpha$$

(approximately, for large enough n), or

$$\mathbf{P}(-\frac{z_{\alpha/2}\sigma}{\sqrt{n}} \leq \overline{X}-\mu \leq \frac{z_{\alpha/2}\sigma}{\sqrt{n}}) = 1-\alpha.$$

It can be rewritten as

$$\mathbf{P}(-\frac{z_{\alpha/2}\sigma}{\sqrt{n}} \leq -\overline{X}+\mu \leq \frac{z_{\alpha/2}\sigma}{\sqrt{n}}) = 1-\alpha.$$

It gives

$$\mathbf{P}(\overline{X}-\frac{z_{\alpha/2}\sigma}{\sqrt{n}} \leq \mu \leq \overline{X}+\frac{z_{\alpha/2}\sigma}{\sqrt{n}}) = 1-\alpha.$$

Therefore, the interval

$$\left[\overline{X}-\frac{z_{\alpha/2}\sigma}{\sqrt{n}}, \overline{X}+\frac{z_{\alpha/2}\sigma}{\sqrt{n}}\right]$$

is $(1-\alpha)100\%$-confidence interval for the true unknown μ.

The case of an unknown population variance

Consider an i.i.d. sample $X_1, ..., X_n$ from a population distribution with unknown mean μ and unknown variance σ^2. In this case, construction of the confidence intervals for μ requires some approximation of σ^2.

- We can calculate sample variance S^2 and simply replace σ^2 by S^2.
- If we use a hypothesis about a one-parameter distribution, where $\sigma = \sigma(\mu)$ is uniquely defined by μ, then the confidence intervals have to be constructed as above, with σ replaced by $\sigma(\overline{\mu}) = \sigma(\overline{X})$.

Confidence intervals for one-parameter distributions

If a distribution is completely defined by a single parameter, then $\sigma = \sigma(\mu)$. For example, if X has Bernoulli distribution with a parameter $p \in (0,1)$, then $\mu = \mathbf{E}X = p$ and $\sigma^2 = \operatorname{Var} X = p(1-p)$. We can replace σ^2 by $\overline{X}(1-\overline{X})$. Another example: if $X \sim Exp(\lambda)$ with a parameter $\lambda > 0$, then $\mu = \mathbf{E}X = 1/\lambda$ and $\sigma^2 = \operatorname{Var} X/\lambda^2$. We can replace σ^2 by $1/\overline{X}^2$.

The construction of the confidence interval for μ for these distributions can be illustrated by the following example.

Example: proportion of the binomial distribution. Consider an election survey, which is based on 100 randomly asked voters, shows that 40 voters supported a party A and 60 voters supported a party B. Let us find 90% confidence interval for the proportion p of the voters supporting B using the normal approximation for the distribution of p.

Let $\hat{p} = 0.6$ be the proportion obtained via survey on n people, with $n = 100$.

Let p be the true unknown proportion. We have that

$$n\hat{p} \sim Bin(n,p), \qquad \mathbf{E}\hat{p} = p, \qquad \operatorname{Var}\hat{p} = p(1-p)/n.$$

The distribution of $\hat{p}$ can be considered approximately as $N(p, p(1-p)/n)$. Since p is unknown, we have to approximate it again by $N(p, \hat{p}(1-\hat{p})/n)$. Therefore, we accept that

$$X = \frac{\hat{p}-p}{\sqrt{\hat{p}(1-\hat{p})n}}$$

is distributed as $N(0,1)$, and that, for $\alpha \in (0,1)$, we have

$$\mathbf{P}\left(-z_{\alpha/2} < X < z_{\alpha/2}\right) = 1-\alpha.$$

For this example, we have $\alpha = 0.1$, and

$$z_{\alpha/2} = 1.65, \quad \sqrt{\hat{p}(1-\hat{p})/n} = 0.049, \quad z_{\alpha/2}\sqrt{\hat{p}(1-\hat{p})/n} = 0.0808.$$

Hence

$$\begin{aligned}
1-\alpha &= \mathbf{P}\left(\hat{p} - z_{\alpha/2}\sqrt{\hat{p}(1-\hat{p})/n} < p < \hat{p} + z_{\alpha/2}\sqrt{\hat{p}(1-\hat{p})/n}\right) \\
&= \mathbf{P}\left(\hat{p} - 0.0808 < p < \hat{p} + 0.0808\right) \\
&= \mathbf{P}\left(0.6 - 0.0808 < p < 0.6 + 0.0808\right) \\
&= \mathbf{P}\left(0.5192 < p < 0.6808\right).
\end{aligned}$$

Therefore, the interval $[0.5192, 0.6808]$ is 90%-confidence interval for the true unknown p.

Confidence intervals for the Gaussian samples

The confidence intervals for the mean and variance of a normal distribution can be constructed explicitly via the critical values of t and χ^2 distributions.

The confidence interval for the mean of a normal distribution. Assume that $X_1, ..., X_n$ are i.i.d. (independent and identically distributed) random variables distributed as $N(\mu, \sigma^2)$, with the sample mean $\overline{X}$. Let us construct $(1-\alpha)\%$-confidence interval for the mean for a given $\alpha \in (0,1)$.

Let $Y = \frac{\overline{X}-\mu}{\sqrt{S^2/n}}$. We have that $Y \sim t_{n-1}$ (the Student's distribution), and

$$\mathbf{P}\left(-t_{n-1,\alpha/2} \leq Y \leq t_{n-1,\alpha/2}\right) = 1-\alpha,$$

where $t_{n-1,\alpha/2}$ and $-t_{n-1,\alpha/2} = t_{n-1,1-\alpha/2}$ are the critical values for the distribution t_{n-1}. Hence

$$\mathbf{P}\left(-t_{n-1,\alpha/2} \leq \frac{\overline{X}-\mu}{\sqrt{S^2/n}} \leq t_{n-1,\alpha/2}\right) = 1-\alpha,$$

and

$$\mathbf{P}\left(\overline{X} - t_{n-1,\alpha/2}\sqrt{S^2}/n \leq \mu \leq \overline{X} + t_{n-1,\alpha/2}\sqrt{S^2}/n\right) = 1-\alpha.$$

Therefore, the interval

$$\left[\overline{X} - t_{n-1,\alpha/2}\sqrt{S^2}/n, \overline{X} + t_{n-1,\alpha/2}\sqrt{S^2}/n\right]$$

is $(1-\alpha)100\%$-confidence interval for the true unknown μ.

The confidence interval for the variance of a normal distribution. Assume that we observe an i.i.d. sample $X_1, ..., X_n$ from a normal distribution with the sample variance S^2. Let us construct $(1-\alpha)\%$-confidence interval for the variance.

We have that $\frac{(n-1)}{\sigma^2}S^2 \sim \chi^2_{n-1}$ and

$$\mathbf{P}\left(\chi^2_{1-\alpha/2} \leq \frac{(n-1)}{\sigma^2}S^2 \leq \chi^2_{\alpha/2}\right) = 1-\alpha.$$

Hence

$$\mathbf{P}\left(1/\chi^2_{\alpha/2} \leq \frac{\sigma^2}{(n-1)S^2} \leq 1/\chi^2_{1-\alpha/2}\right) = 1-\alpha,$$

and

$$\mathbf{P}\left(\frac{(n-1)S^2}{\chi^2_{\alpha/2}} \leq \sigma^2 \leq \frac{(n-1)S^2}{\chi^2_{1-\alpha/2}}\right) = 1-\alpha.$$

Therefore, the interval

$$\left[\frac{(n-1)S^2}{\chi^2_{\alpha/2}}, \frac{(n-1)S^2}{\chi^2_{1-\alpha/2}}\right]$$

is $(1-\alpha)100\%$-confidence interval for the true unknown σ^2.

12.3 Hypothesis testing

Null hypothesis and alternative hypothesis

To analyze the reliability of the estimate, there is a special approach named *hypothesis testing*. In this framework, two hypotheses about the probability distribution of the underlying sample are selected; one is called the null hypothesis and is denoted by H_0 and the other is called the alternative hypothesis and is denoted by H_A. (There are several common variations on this notation.)

There is an asymmetry between the null and alternative hypotheses, as can be seen later.

Definition 12.1 *If a hypothesis completely specifies the probability distribution then such a hypothesis is called simple hypothesis. Otherwise, it is called composite hypothesis.*

H_0 might state that the distribution was $N(\mu_0, \sigma^2)$ (a simple hypothesis) and H_A might state:

- that the distribution was $N(\mu_a, \sigma^2)$, for some given μ_a (a simple hypothesis);
- that the distribution was $N(\mu, \sigma^2)$, for some unknown $\mu \neq \mu_0$ (a composite hypothesis);
- that the distribution was not $N(\mu_0, \sigma^2)$ (a composite hypothesis).

Hypothesis that $\mu > 0.2$ or that $\mu \neq 0.25$ are, respectively, examples of one-sided and two-sided alternative hypotheses.

The null hypothesis might state that the distribution is Poisson with an unknown mean, and the alternative hypothesis might state that the distribution is not Poisson. Neither of these hypotheses completely specify the probability distribution. They are examples of composite hypotheses.

The types of errors

We introduce below the so-called Neyman-Pearson paradigm commonly used in hypothesis testing. Under this paradigm, a decision whether to or not to reject H_0 in favor of H_A is made on the basis of $T(X)$, where X denotes the sample values and $T(X)$ is a statistic, i.e., some value calculated as a function of the sample value $X = (X_1, ..., X_n)$.

Assume that certain algorithm of decision making based on a statistic is selected. For every particular algorithm, the following errors are possible.

Definition 12.2 *If H_0 is rejected when it is true, then an error is called a type I error and its probability is denoted by α.*

If H_0 is simple, α is called the significance level of the test.

Definition 12.3 *If H_0 is accepted when it is false, then an error is called a type II error and its probability is denoted by β. The probability $1-\beta$ that H_0 is rejected when it is false is called the power of the test.*

If H_A is composite, the probability of a type II error will generally depend on which particular member of H_A holds.

Example 12.4 Assume the H_0 hypothesis is $X \sim U(0,1)$ and the H_A hypothesis is $X \sim U(0,2)$, where we have a single measurement X only. Let us accept the following test (or rule): if $X < 0.9$ then we accept H_0. Then

$$\alpha = \mathbf{P}_0(X > 0.9) = 0.1$$

and

$$\beta = \mathbf{P}_A(X < 0.9) = \frac{0.9}{2} = 0.45.$$

Here $\mathbf{P}_0$ and $\mathbf{P}_A$ are the probabilities generated by the hypothesis H_0 or H_A respectively. Therefore, the power of this test is $1 - 0.45 = 0.55$.

Example 12.5 Assume the H_0 hypothesis is $X \sim U(0,1)$ and the H_A hypothesis is $X \sim U(0,1.1)$, where we have a single measurement X only. Let us accept again the following test (or rule): if $X < 0.9$ then we accept H_0. Then

$$\alpha = \mathbf{P}_0(X > 0.9) = 0.1$$

and

$$\beta = \mathbf{P}_A(X < 0.9) = \frac{0.9}{1.1} \sim 0.8.$$

Therefore, the power of this test is $1 - 0.8 = 0.2$.

On the choice of H_0 and H_A

It is conventional to choose the simpler of two hypotheses as the null, in which case it is easier to calculate α.

The consequences of incorrectly rejecting one hypothesis may be greater than those of incorrectly rejecting the other. In such a case, the former should be chosen as the null hypothesis, since the probability of falsely rejecting it could be controlled by choosing α.

Acceptance and rejection over test statistic

Let $T(X)$ be some statistic calculated from a sample X (for instance, the sample mean).

Definition 12.6 *The set of values of $T(X)$ for which H_0 is accepted is called the acceptance region of the test.*

Definition 12.7 *The set of values of $T(X)$ for which H_0 is rejected is called the rejection region of the test.*

Definition 12.8 *If the rejection region is of the form $\{T > c\}$ or $\{T < c\}$, where T is the test statistic, the number c is called the critical value of the test; the critical value separates the rejection region and the acceptance region.*

The p-value for a test statistic

Let $\mathbf{P}_0$ be the probability under H_0. Suppose that the rejection region is $\{T(X) > c\}$ for the probability α of type I error, i.e. $\mathbf{P}_0(T(X) > c) = \alpha$. Assume that the observed statistics has the value $\hat{T}$. The value $\mathbf{P}_0(T(X) \geq \hat{T})$ is called *p-value.* Sometimes, it is more convenient to use this value instead of rejection region. If p-value is less than α, then this means that the null hypothesis has to be rejected, with the probability α of type I error. It can be observed that $\hat{T} > c$ if and only if p-value is less than α. Similarly, we can consider the case where the rejection region is $\{T(X) < c\}$ for the probability α of type I error, and $\mathbf{P}_0(T(X) < c) = \alpha$. Suppose that the observed statistics has the value $\hat{T}$. The value $\mathbf{P}_0(T(X) \leq \hat{T})$ is called p-value again. If p-value is less than α, then this means that the null hypothesis can be rejected, with the probability α of type I error. It can be observed that $\hat{T} < c$ if and only if p-value is less then α.

Calculation of the rejection region based on the Gaussian approximation

In many cases, the exact distribution of a statistics is unknown, and, therefore, not possible to calculate the exact value of α. We could use approximations such as normal approximation for $\sum_{i=1}^{n} X_i$ from the Central Limit Theorem. This can be illustrate by the following example.

Let $X_1, ..., X_n$ be a random sample from an exponential distribution with the parameter λ. Assume that H_0 hypothesis is that $\mathbf{E}X_i = 1/2$, i.e., $\lambda = 2$ and that H_A hypothesis is that $\mathbf{E}X_i = 1$, i.e., $\lambda = 1$. Let $\mathbf{P}_0 = \mathbf{P}(\cdot|\lambda = 2)$ be the probability under H_0. We will reject H_0 if, for some c,

$$\sum_{i=1}^{n} X_i > c.$$

Let us find c that ensures the probability of type I error is less or equal $\alpha = 0.05$. It suffices to find c such that

$$\alpha = 0.05 = \mathbf{P}_0\left(\sum_{i=1}^{n} X_i > c\right) = \mathbf{P}\left(\sum_{i=1}^{n} X_i > c \,\Big|\, \lambda = 2\right).$$

We have $\mathbf{E}(X_i|\lambda = 2) = 1/2$ and $\mathrm{Var}\,(X_i|\lambda = 2) = 1/4$.

Let $W = \sum_{i=1}^{n} X_i = \sum_{i=1}^{n} X_i$. We have that, under H_0, $\mathbf{E}(W|\lambda = 2) = n/2$, $\mathrm{Var}\,(W|\lambda = 2) = n/4$,

By the Central Limit Theorem, we accept that $\frac{W-n/2}{\sqrt{n}/2} \sim N(0,1)$, approximately, for large n. We obtain from the tables that

$$\alpha = \mathbf{P}_0\left(\sum_{i=1}^{n} X_i > c\right) = \mathbf{P}_0\left(\frac{W - n/2}{\sqrt{n}/2} > \frac{c - n/2}{\sqrt{n}/2}\right) = 1 - \Phi\left(\frac{c - n/2}{\sqrt{n}/2}\right).$$

It gives $(2c - n)/\sqrt{n} = q_{1-\alpha}$, i.e., the $(1 - \alpha)$th quantile of $N(0,1)$.

For $\alpha = 0.05$, it gives $q_{0.95} = 1.65$. For $n = 20$, it gives $(2c - 20)/\sqrt{n} = 1.6$, or $c - 10 = \frac{1.65}{2}\sqrt{20}$, or

$$c = 10 + \frac{1.65}{2}\sqrt{20}.$$

The rejection region is

$$\{x \in \mathbf{R}^n : \sum_{i=1}^{n} x_i > c\}.$$

Further, let us assume that we observed a sample such that $\sum_{i=1}^{n} X_i = 60$, $n = 100$. The corresponding p-value is

$$\mathbf{P}_0\left(\sum_{i=1}^{n} X_i \geq 60\right) = \mathbf{P}_0\left(\frac{W - n/2}{\sqrt{n}/2} \geq \frac{60 - n/2}{\sqrt{n}/2}\right)$$
$$= \mathbf{P}_0\left(\frac{W - 50}{5} \geq \frac{10}{5}\right) = \mathbf{P}\,(Z \geq 2) > 1 - \alpha,$$

for $Z \sim N(0,1)$.

Example: Hypothesis testing for the ratio of two variances

Consider two samples, one with 13 independent members, and another with 11 independent members, and with the sample variances $S_1^2 = 2$ and $S_2^2 = 5$, respectively. We assume that the samples are from the distributions $N(a_1, \sigma_1^2)$ and $N(a_2, \sigma_2^2)$ respectively.

Consider two hypotheses:

- H_0: $\sigma_1^2 = \sigma_1^2$, i.e., the variances for these two distributions are the same, and
- H_A: $\sigma_1^2 \neq \sigma_1^2$, i.e., the variances for these two distributions are different.

Let us construct the acceptance interval and test these hypotheses with 90% level of confidence. We have that

$$\frac{\sigma_2^2 S_1^2}{\sigma_1^2 S_2^2} \sim F_{12,10}.$$

Hence, under H_0,

$$\frac{S_1^2}{S_2^2} \sim F_{12,10}.$$

The test statistics is $F = S_1^2/S_2^2 = 0.4$. We reject H_0 if $F \notin (f_{0.95}, f_{0.05})$, where

$$f_{0.05} = 2.913, \quad f_{0.95} = 1/f_{0.05}^{10,12} = 1/2.753 = 0.36$$

are the corresponding critical values for the distribution $F_{12,10}$, and where $f_{0.05}^{10,12} = 2.753$ is the corresponding critical value for the distribution $F_{10,12}$. Since $F \notin (f_{0.95}, f_{0.05})$, we reject H_0 with 90% level of confidence.

12.4 Confidence intervals and hypothesis testing

There is a duality between confidence intervals (more generally, confidence sets) and hypothesis tests. In this section, we will show that a confidence set can be obtained by "inverting" a hypothesis test, and vice versa.

Let θ be a parameter of a family of probability distributions, and denote the set of all possible values of θ by Θ. Denote the random variables constituting the data by X.

Theorem 12.9 *Suppose that for every value $\theta \in \Theta$ there is a test at level α of the hypothesis H_0 that $\theta = \theta_0$. Denote the acceptance region of the test by $A(\theta_0)$. Then the set*

$$C(X) = \{\theta : X \in A(\theta)\}$$

is a $100(1-\alpha)\%$ confidence region for θ.

In words, the above theorem says that $100(1-\alpha)\%$ confidence region for θ consists of those values of θ for which the hypothesis that $\theta = \theta_0$ will not be rejected at level α.

Proof. Since A is the acceptance region of a test at level α, we have

$$\mathbf{P}(X \in A(\theta_0)|\theta = \theta_0) = 1 - \alpha.$$

By the definition of $C(X)$, we obtain

$$\mathbf{P}(\theta_0 \in C(X)|\theta = \theta_0) = \mathbf{P}(X \in A(\theta_0)|\theta = \theta_0) = 1 - \alpha.$$

Theorem 12.10 *Assume that $C(X)$ is a $100(1-\alpha)\%$ confidence region for $\theta \in \Theta$. Then an acceptance region for a test at level α of the hypothesis H_0 that $\theta = \theta_0$ is*

$$A(\theta_0) = \{X : \theta_0 \in C(X)\}.$$

In other words, the above theorem says that the hypothesis that $\theta = \theta_0$ is accepted if θ_0 lies in the confidence region.

Proof. We have that, for every θ,

$$\mathbf{P}(\theta \in C(X)|\theta) = 1 - \alpha.$$

This acceptance region gives a test at level α since

$$\mathbf{P}(X \in A(\theta_0)|\theta = \theta_0) = \mathbf{P}(\theta_0 \in C(X)|\theta = \theta_0) = 1 - \alpha.$$

Problems for Week 12

Problem 12.1 *A proportion of the terms with certain property in a sample of the size 250 is found to be 0.5. Find 90% confidence interval for the proportion of the entire population with this property.*

Problem 12.2 *The sample variance S was calculated for the sample of 9 independent random variables with the same normal distribution $N(\mu, \sigma^2)$. Find the 90%-confidence interval for σ^2 given that $S^2 = 3$.*

Problem 12.3 *A proportion of the terms with certain property in a sample of the size 100 is found to be 0.4. Let p be the proportion of the entire population with this property. Consider two hypotheses:*

- *H_0: $p = 0.5$, and*
- *H_A: $p < 0.5$.*

Test the hypotheses with the significance level $\alpha = 0.05$ (i.e., with the probability of the type I error $\alpha = 0.05$). Use the p-value.

Problem 12.4 *Consider two samples with 10 independent members each, and with sample variances $S_1^2 = 4$ and $S_2^2 = 2$, respectively. We assume that the samples are from the distributions $N(a_1, \sigma_1^2)$ and $N(a_1, \sigma_1^2)$ respectively. Consider two hypotheses:*

- *H_0: the variances for two distributions are the same, and*
- *H_A: the variances for two distributions are different.*

Test these hypotheses with 90% level of confidence.

Problem 12.5 *Consider two samples, one with 11 independent members, and another with 13 independent members, and with the sample variances $S_1^2 = 2$ and $S_2^2 = 4$, respectively. We assume that the samples are from the distributions $N(a_1, \sigma_1^2)$ and $N(a_2, \sigma_2^2)$ respectively.*

Consider two hypotheses:

- *H_0: $\sigma_1^2 = \sigma_2^2$, i.e., the variances for these two distributions are the same, and*
- *H_A: $\sigma_1^2 \neq \sigma_2^2$, i.e., the variances for these two distributions are different.*

Construct the acceptance interval and test these hypotheses with 90% level of confidence.

Problem 12.6 *Let $X_1, ..., X_n$ be a random sample from $N(\mu, \sigma^2)$ given some σ. Consider two hypotheses:*

- *H_0: $\mu = 2$, and*
- *H_A: $\mu = 1$.*

We reject H_0 if the sample mean is too small. For $\sigma = 1$ and $n = 20$, find the rejection region for for 95% level of confidence (i.e., for the probability of the type I error $\alpha = 0.05$).

Problem 12.7 *In the framework of Problem 12.6, for a general $\mu > 2$, find the acceptance region $A(\mu)$ for 95% level of confidence. Assume that $\mu_A = 1$ and that we know that the test leads to rejection of H_0 if $\sum_i X_i < c$, for some $c = c(\mu)$. Find the confidence region for μ given a sample.*

Appendix 1: Solutions for the Problems for Weeks 1-12

In this appendix, the problems are listed such that their solutions do not require knowledge of the consequent chapters. Therefore, they can be used for tutorials.

Solutions

Solutions for problems for Week 1

Solution for Problem 1.1. We have $\mathbf{P}(A \cap B) = 0.5 \cdot 0.4 = 0.2$.

Solution for Problem 1.2. We have $\mathbf{P}(A \cup B) = \mathbf{P}(A) + \mathbf{P}(B) - \mathbf{P}(A \cap B) = 0.8 + 0.4 - 0.8 \cdot 0.4 = 0.88$.

Solution for Problem 1.3. From Venn diagram, we can see that

$$\begin{aligned}\mathbf{P}(A \cap B^c) &= \mathbf{P}(A \cup B) - \mathbf{P}(B) = \mathbf{P}(A) + \mathbf{P}(B) - \mathbf{P}(A \cap B) - \mathbf{P}(B)\\ &= \mathbf{P}(A) + \mathbf{P}(B) - \mathbf{P}(A)\mathbf{P}(B) - \mathbf{P}(B) = \mathbf{P}(A) - \mathbf{P}(A)\mathbf{P}(B)\\ &= \mathbf{P}(A)(1 - \mathbf{P}(B)) = \mathbf{P}(A)\mathbf{P}(B^c).\end{aligned}$$

By the definition, it follows that the events A and B^c are independent.

Solution for Problem 1.4. Observe that the events A and B^c are independent and $\mathbf{P}(B^c) = 1 - 0.4 = 0.6$. Hence $\mathbf{P}(A \cup B^c) = 0.5 \cdot 0.6 = 0.3$.

Solution for Problem 1.5. We have $\mathbf{P}(A \backslash B) = \mathbf{P}(A \cap B^c) = \mathbf{P}(A)\mathbf{P}(B^c) = 0.8 \cdot 0.6 = 0.48$.

Solution for Problem 1.6. Consider events $D = \{\text{detail is defective}\}$, $A = \{\text{from machine A}\}$, $B = \{\text{from machine B}\}$. By Bayes' formula,

$$\mathbf{P}(D) = \mathbf{P}(D|A)\mathbf{P}(A) + \mathbf{P}(D|B)\mathbf{P}(B) = 0.1 \times 0.4 + 0.05 \times 0.6$$
$$= 0.04 + 0.03 = 0.07.$$

By Bayes' formula again, $\mathbf{P}(B|D) = \frac{\mathbf{P}(D|B)\mathbf{P}(B)}{\mathbf{P}(D)} = \frac{0.05\times 0.6}{0.07} = 3/7$.

Solutions for problems for Week 2

Solution for Problem 2.1. Since $0.2+0.3+0.4+c = 1$, we have that $c = 0.1$.

Solution for Problem 2.2. We have that $F(-1) = 0$, $F(1) = 0.1$, $F(4.5) = 0.8$.

Solution for Problem 2.3. We have that $\mathbf{P}(1 < X \le 4.5) = F(4.5) - F(1) = 0.7$.

Solution for Problem 2.4. Let X be the number trials where the red comes up. We have that

$$F(0.5) = \mathbf{P}(X \le 0.5) = \mathbf{P}(X = 0) = \left(\frac{20}{38}\right)^5.$$

Solution for Problem 2.5. We have that

$$\int_{-\infty}^{\infty} f(x)dx = 1.$$

Hence

$$\int_{-\infty}^{\infty} f(x)dx = \int_0^4 kxdx = k\int_0^4 xdx = k\frac{x^2}{2}\Big|_0^4 = 16k/2 = 1.$$

Hence $k = 1/8$.

Solution for Problem 2.6. We have

$$\mathbf{P}(-2 < X < 1) = \int_{-2}^{1} f(x)dx = \int_0^1 f(x)dx = \frac{1}{8}\int_0^1 xdx = \frac{1}{8}\frac{x^2}{2}\Big|_0^1 = 1/16.$$

Solution for Problem 2.7. The density $f(x)$ for X is $f(x) = \frac{1}{4}\mathbb{I}_{[0,4]}(x)$. Further,

$$\mathbf{P}(X^2 < 2) = \mathbf{P}(X \in (0, \sqrt{2}) = \int_0^{\sqrt{2}} f(x)dx = \int_0^{\sqrt{2}} \frac{1}{4}dx = \frac{1}{4}x\Big|_0^{\sqrt{2}} = \sqrt{2}/4.$$

Solution for Problem 2.8. Probability of a success in one game is $p = 1/38$. We use Poisson distribution with $\lambda = 38 \cdot p = 1$. Let Y be the number of occurrence of 1 in 3 games. We have

$$\mathbf{P}(Y \geq 1) = 1 - \mathbf{P}(Y = 0) = 1 - e^{-1} = 0.632.$$

Solution for Problem 2.9. Probability of a success in one game is $p = 1/38$. We use the Poisson distribution with $\lambda = 76 \cdot p = 2$. Let Y be the number of occurrence of 1 in 76 games. We have

$$\mathbf{P}(Y = 0) = e^{-2} = 0.1353,$$
$$\mathbf{P}(Y \geq 1) = 1 - \mathbf{P}(Y = 0) = 1 - e^{-2} = 0.8647.$$

Solution for Problem 2.10. (i) We observe that $\mathbf{P}(Y = 0.7) > 0$ and that Y can take any value in a continuous interval $[0, 0.7]$. Hence the distribution of Y is neither continuous nor discrete. Furthermore, we have that the c.d.f. is $F_Y(y) = \mathbf{P}(Y \leq y) = 0$ if $y < 0$ and $F_Y(y) = \mathbf{P}(Y \leq y) = \mathbf{P}(X \leq y) = y$ if $y \in [0, 0.7]$. Finally, $F_Y(y) = \mathbf{P}(Y \leq y) = 1$ if $y \geq 0.7$. It can be noted that the function F_Y is neither continuous nor piecewise constant; this shows again that it is neither continuous nor discrete distribution.

(ii) We have that the c.d.f. for W is

$$\begin{aligned} F_W(w) &= \mathbf{P}(W \leq w) = \mathbf{P}(2 - 2X \leq w) = \mathbf{P}(1 - X \leq w/2) \\ &= \mathbf{P}(X \geq 1 - w/2) = 1 - \mathbf{P}(X < 1 - w/2) = 1 - F_X(1 - w/2), \end{aligned}$$

where

$$F_X(x) = \begin{cases} 0, & x < 0, \\ x, & x \in (0,1) \\ 1, & x > 1. \end{cases}$$

Hence

$$F_W(w) = \begin{cases} 1, & 1 - w/2 < 0, \\ w/2, & 1 - w/2 \in (0,1) \\ 0, & 1 - w/2 > 1 \end{cases} = \begin{cases} 1, & w > 2, \\ w/2, & w \in (0,2) \\ 0, & w < 0. \end{cases}$$

It can be seen that W has a uniform distribution again.

We have that the c.d.f. for Z is

$$F_Z(z) = \mathbf{P}(Z \le z) = \mathbf{P}(3Y + 2 \le z) = \mathbf{P}(Y \le (z-2)/3)$$
$$= F_Y((z-2)/3),$$

where

$$F_Y(y) = \begin{cases} 0, & y < 0, \\ y, & y \in (0, 0.7) \\ 1, & y \ge 0.7. \end{cases}$$

Hence

$$F_Z(z) = \begin{cases} 0, & (z-2)/3 < 0, \\ (z-2)/3, & (z-2)/3 \in (0, 0.7) \\ 1, & (z-2)/3 \ge 0.7 \end{cases} = \begin{cases} 0, & z < 2, \\ (z-2)/3, & z \in (2, 4.1) \\ 1, & z \ge 4.1. \end{cases}$$

Solution for Problem 2.11. Let X be the random number of emails arriving during any one hour. We have that $X \sim Poi(\lambda)$, and $\mathbf{E}X = \lambda = 2$. Let Y be the random number of emails arriving during any two hours. We have that $Y \sim Poi(2\lambda)$, and $\mathbf{E}Y = 2\lambda = 4$, and $Y \sim Poi(4)$. It gives that

$$P(Y = 0) = \frac{4^0}{0!}e^{-4}.$$

Solutions for problems for Week 3

Solution for Problem 3.1. k can be found from the equation

$$\int_{\mathbf{R}^2} f(x,y)dxdy = 1,$$

or

$$\int_D f(x,y)dxdy = k\int_D ydxdy = k\int_{-1}^1 dx\frac{1}{2} = \frac{k}{2}\int_{-1}^1 dx = \frac{k}{2} = 1.$$

Hence $k = 2$.

Further,

$$f_X(x) = \int_{-\infty}^{\infty} f(x,y)dy = \int_0^1 2ydy = 0.5 \times 2 = 1, \quad x \in [-1,1],$$
$$f_X(x) = 0, \quad x \notin [-1,1].$$

Similarly,

$$f_Y(y) = \int_{-\infty}^{\infty} f(x,y)dx = \int_0^1 2ydx = 2y, \quad y \in [0,1],$$
$$f_Y(y) = 0, \quad y \notin [0,1].$$

We have

$$\mathbf{P}(Y < 1/2) = \int_0^{1/2} f_Y(y)dy = \int_0^{1/2} 2ydy = 2\frac{y^2}{2}|_0^{1/2} = 1/4.$$

Solution for Problem 3.2. The value k can be found from the equation

$$\int_{\mathbf{R}^2} f(x,y)dxdy = 1,$$

or

$$\int_D f(x,y)dxdy = k\int_D ydxdy = k\int_0^1 dx \int_{-1}^0 ydy = k\int_0^1 dx \frac{y^2}{2}\bigg|_{-1}^0$$
$$= \frac{k}{2}\int_0^1 dx = -\frac{k}{2} = 1.$$

Hence $k = -2$.

Further,

$$f_Y(y) = \int_{-\infty}^{\infty} f(x,y)dx = -\int_0^1 2ydx = -2y, \quad y \in [-1,0],$$
$$f_Y(y) = 0, \quad y \notin [-1,0].$$

We have

$$\mathbf{P}(Y < -1/2) = \int_{-1}^{-1/2} f_Y(y)dy = \int_{-1}^{-1/2} 2ydy = -2\frac{y^2}{2}|_{-1}^{-1/2} = 3/4.$$

Solution for Problem 3.3. By Bayes's formula,

$$\mathbf{P}(Y + |X| < 1) = \mathbf{P}(Y + |X| < 1|Y = 0.3)\mathbf{P}(Y = 0.3)$$
$$+\mathbf{P}(Y + |X| < 1|Y = 0.7)\mathbf{P}(Y = 0.7).$$

It gives

$$\mathbf{P}(Y + |X| < 1) = \mathbf{P}(0.3 + |X| < 1|Y = 0.3)\frac{1}{5} + \mathbf{P}(0.7 + |X| < 1|Y = 0.7)\frac{4}{5}$$
$$= \mathbf{P}(|X| < 0.7|Y = 0.3)\frac{1}{5} + \mathbf{P}(|X| < 0.3|Y = 0.7)\frac{4}{5}.$$

By independence of Y and Y, it can be simplified as

$$\mathbf{P}(Y+|X|<1)=\mathbf{P}(|X|<0.7)\frac{1}{5}+\mathbf{P}(|X|<0.3)\frac{4}{5}=0.7\frac{1}{5}+0.3\frac{4}{5}=0.38.$$

We use here that $\mathbf{P}(|X|<a)=\frac{2a}{2}=a$, for $a=0.3$ and $a=0.7$.

Solution for Problem 3.4. (a) We have the joint density $f(x,y)=c\mathbb{I}_D(x,y)$, where

$$D=\{(x,y): x\in[0,1],\quad y\in[0,1-x]\}.$$

The constant c can be found from the equation

$$\int_{\mathbf{R}^2} f(x,y)dxdy=c\int_D dxdy=1.$$

We have

$$\int_D dxdy=\int_0^1 dx\int_0^{1-x}dy=\int_0^1(1-x)dx=(x-x^2/2)\Big|_0^1=1/2.$$

Hence $c=2$.

Further,

$$f_X(x)=c\int_{-\infty}^{\infty}f(x,y)dy=2\int_0^{1-x}dy=2(1-x),\quad x\in[0,1],$$
$$f_X(x)=0,\quad x\notin[0,1],$$

$$f_Y(y)=\int_{-\infty}^{\infty}f(x,y)dx=\int_0^{1-y}f(x,y)dx=2(1-y),\quad y\in(0,1),$$
$$f_Y(y)=0,\quad y\notin[0,1].$$

We have

$$f_{X|Y}(x|y)=\frac{f(x,y)}{f_Y(y)}=\frac{2\mathbb{I}_{D)}}{f_Y(y)},$$
$$f_{Y|X}(y|x)=\frac{f(x,y)}{f_X(x)}=\frac{2\mathbb{I}_{D)}}{f_X(x)}.$$

(b)

$$\mathbf{P}(X+Y<0.5)=\int_0^{0.5}dx\int_0^{0.5-x}f(x,y)dy=\int_0^{0.5}dx\int_0^{0.5-x}2dy$$
$$=\int_0^{0.5}dx2(0.5-x)=(x-x^2)\Big|_0^{0.5}=0.5-0.25=0.25.$$

(c) Similarly,

$$\mathbf{P}(X+Y>0.5)=\int_0^{0.5}dx\int_0^{1-x}f(x,y)dy=\int_0^{0.5}dx\int_0^{1-x}2dy$$
$$=\int_0^{0.5}dx2(1-x)=(2x-x^2)\Big|_0^{0.5}=1-0.25=0.75.$$

(d) We have that $\mathbf{P}(0<Y<1-X)=1$, hence $\mathbf{P}(Y>1-X)=0$. Hence

$$\mathbf{P}(Y>1-0.5|X=0.5)=0.$$

We can use also the fact that $f_Y(y|x)=0$ for $y>0.5$ if $x=0.5$.

Solution for Problem 3.5. (a) We look for c from the equation

$$\int_1^{\infty}f_X(x)dx=c\int_1^{\infty}e^{-2x}dx=1.$$
$$\int_1^{\infty}e^{-2x}dx=\frac{1}{2}e^{-2}.$$

Hence $c=2e^2=14.78$.

(b) We look for $M>1$ such that

$$\int_1^M f_X(x)dx=0.5.$$

It gives

$$\int_1^M 2e^2e^{-2x}dx=0.5,$$

or

$$2\frac{1}{-2}e^{2-2x}\Big|_1^M=1-e^{2-2M}=0.5,$$

or $e^{2-2M}=0.5$. Hence $2-2M=\ln(1)-ln(2)$, or $M=(\ln(2)+2)/2=1.3466$.

(c) Y has the density $f_Y(y)=0.5\mathbb{I}_{[0,2]}(y)$, and

$$f(x,y)=f_X(x)f_Y(y)=0.5\mathbb{I}_{[0,2]}(y)2e^2e^{-2x}\mathbb{I}_{[1,+\infty)}(x)$$
$$=e^2e^{-2x}\mathbb{I}_{[1,+\infty)\times[0,2]}(x,y).$$

(d)

$$\mathbf{P}(X<Y)=\int_1^2dy\int_1^y f(x,y)dx=\int_1^2dy\int_1^y e^{2(1-x)}dy$$
$$=e^2\int_1^2dy\frac{1}{-2}e^{-2x}\Big|_1^y=\frac{1}{2}e^2\int_1^2dy(e^{-2}-e^{-2y})=0.2835.$$

Solutions for problems for Week 4

Solution for Problem 4.1. Let $g(x) = -x/2$ and $h(u) = -2u$. The Jacobian is

$$J(u) = \frac{dh}{du}(u) = -2.$$

Therefore, the density for U is

$$f_U(u) = f_X(h(u))|J(u)| = f_X(h(u))2 = 2\lambda \exp[-\lambda(-2u)]\mathbb{I}_{\{u<0\}}(u).$$

Solution for Problem 4.2. By defining $g(x) = -x/2 + 3$ and $h(u) = -2u + 6$, the Jacobian is

$$J(u) = \frac{dh}{du}(u) = -2.$$

Hence, the density for U is

$$f_U(u) = f_X(h(u))|J(u)| = f_X(h(u))2 = \lambda \exp[-\lambda(-2u+6)]\mathbb{I}_{\{-2u+6>0\}}(u).$$

Solution for Problem 4.3. For $y \leq 0$, we have $F_Y(y) = 0$. For $y > 0$, we have

$$\begin{aligned} F_Y(y) = \mathbf{P}(Y \leq y) = \mathbf{P}(X^2 \leq y) &= \mathbf{P}(-y^{1/2} \leq X \leq y^{1/2}) \\ &= F_X(y^{1/2}) - F_X(-y^{1/2}). \end{aligned}$$

By differentiating, we obtain the corresponding densities. For $y \leq 0$, we have $f_Y(y) = 0$. For $y > 0$, we obtain

$$\begin{aligned} f_Y(y) = \frac{d}{dy}F_Y(y) &= \frac{d}{dy}\{F_X(y^{1/2}) - F_X(-y^{1/2})\} \\ &= f_X(y^{1/2})\frac{d}{dy}y^{1/2} - f_X(-y^{1/2})\frac{d}{dy}(-y^{1/2}) \\ &= f_X(y^{1/2})\frac{y^{-1/2}}{2} + f_X(-y^{1/2})\frac{y^{-1/2}}{2}. \end{aligned}$$

Solution for Problem 4.4. (a) We have $u = g_1(x,y) = x + 2y$, $v = g_2(x,y) = x + y$. It gives

$$X = h_1(U,V) = -U + 2U, \quad Y = h_2(U,V) = U - V.$$

The Jacobian is

$$J(u,v) = \det\begin{pmatrix} -1 & 2 \\ 1 & -1 \end{pmatrix} = 1 \cdot 1 - 1 \cdot 2 = -1.$$

The joint density for U and V is

$$f_{UV}(u,v) = f_{XY}(h_1(u,v), h_2(u,v))|J(u,v)| = f_{XY}(2v-u, u-v)$$
$$= \alpha\beta \exp[-\alpha(2v-u) - \beta(u-v)]\mathbb{I}_{\{v<u<2v\}}(u,v).$$

(Note that the existence of u such that $v < u < 2v$ implies that $v > 0$.)

(b) The marginal density for $V = X + Y$ is

$$f_V(v) = \int_{-\infty}^{\infty} f_{UV}(u,v)du = \int_v^{2v} \alpha\beta \exp[-\alpha(2v-u) - \beta(u-v)]du.$$

(c) If $\alpha \neq \beta$ then

$$f_V(v) = \int_v^{2v} \alpha\beta \exp[-\alpha(2v-u) - \beta(u-v)]du$$
$$= \alpha\beta \exp[-2\alpha v + \beta v)] \int_v^{2v} \exp[\alpha u - \beta u]du.$$

(d) If $\alpha = \beta$ then

$$\int_v^{2v} \exp[\alpha u - \beta u]du = \int_v^{2v} 1 \cdot du = v$$

and

$$f_V(v) = \alpha\beta \exp[-2\alpha v + \beta v)]v = \alpha^2 \exp[-\alpha v)]v.$$

This is a Gamma density.

Solution for Problem 4.5. First, we note that

$$U \leq u \quad \text{if and only if} \quad X \leq u, \quad Y \leq U.$$

Thus,

$$F_U(u) = \mathbf{P}(U \leq u) = \mathbf{P}(X \leq u, Y \leq u).$$

By the assumption about independence, we obtain

$$F_U(u) = \mathbf{P}(X \leq u)\mathbf{P}(Y \leq u) = F_X(u)F_Y(u).$$

Hence

$$F_U(u) = 0, \quad u < 0,$$
$$F_U(u) = \mathbf{P}(X \leq u)\mathbf{P}(Y \leq u) = F_Y(u) = 1 - e^{-\alpha u}, \quad u > 1,$$
$$F_U(u) = \mathbf{P}(X \leq u)\mathbf{P}(Y \leq u) = u(1 - e^{-\alpha u}), \quad u \in (0,1).$$

Differentiating, we find the density

$$f_U(u) = \frac{d}{du}F(u) = 0, \quad x < 0,$$
$$f_U(u) = \frac{d}{du}F(u) = 1 - e^{-\alpha u} + u\alpha e^{-\alpha u}, \quad u \in (0,1),$$
$$f_U(u) = \frac{d}{du}F(u) = \alpha e^{-\alpha u}, \quad u > 1.$$

Solutions for problems for Week 5

Solution for Problem 5.1. We have that $\mathbf{E}X = 1 \cdot 0.1 + 3 \cdot 0.2 + 4 \cdot 0.5 + 5 \cdot 0.2 = 3.7$, $\mathbf{E}X^2 = 1^2 \cdot 0.1 + 3^2 \cdot 0.2 + 4^2 \cdot 0.5 + 5^2 \cdot 0.2 = 14.9$, $\mathrm{Var}\,(X) = EX^2 - (\mathbf{E}X)^2 = 14.9 - 3.7^2 = 1.21$.

Solution for Problem 5.2. We must have $0.2 + 0.3 + 0.2 + k = 1$, i.e., $k = 0.3$. Hence $\mathbf{E}X = 1 \cdot 0.2 + 3 \cdot 0.3 + 4 \cdot 0.3 + 5 \cdot 0.2 = 3.3$.

Solution for Problem 5.3. We have

$$\mathbf{E}X = \int_{-\infty}^{\infty} xf(x)dx = \int_0^4 x \cdot x/8dx = \frac{1}{8}\int_0^4 x^2 dx = \frac{1}{8}\frac{x^3}{3}\Big|_0^4 = 64/24 = 8/3.$$

Further, we have

$$\mathbf{E}X^2 = \int_{-\infty}^{\infty} x^2 f(x)dx = \int_0^4 x^2 \cdot x/8dx = \frac{1}{8}\int_0^4 x^3 dx = \frac{1}{8}\frac{x^4}{4}\Big|_0^4 = 256/32.$$

It gives $VarX = EX^2 - (\mathbf{E}X)^2 = 256/32 - (8/3)^2 = 0.8889$.

Solution for Problem 5.4. We have $f(x) = \mathbb{I}_{[2,8]}(x)a(x-2)$, for some $a \in \mathbf{R}$. We obtain that $a = 1/18$, $h = 1/3$ using that $\int_{-\infty}^{\infty} f(x)dx = \int_2^8 f(x)dx = 1$. It gives

$$\mathbf{E}X = \int_2^8 xf(x)dx = 6, \quad \mathbf{E}X^2 = \int_2^8 x^2 f(x)dx = 38, \quad \mathrm{Var}\,X = 2,$$

$$\mathbf{P}(X < 0) = 0,$$

$$\mathbf{P}(X = 8) = \int_8^8 f(x)dx = 0, \quad \mathbf{P}(X < 5) = \int_2^5 f(x)dx = 1/4.$$

Solution for Problem 5.5. We have

$$\mathbf{E}X = \sum_{k\geq 0} kp(k) = \sum_{k\geq 0} k\frac{\lambda^k}{k!}e^{-\lambda} = \sum_{k\geq 1}\frac{\lambda^k}{(k-1)!}e^{-\lambda} = \lambda\sum_{k\geq 1}\frac{\lambda^{k-1}}{(k-1)!}e^{-\lambda} = \lambda\sum_{m\geq 0}\frac{\lambda^m}{m!}e^{-\lambda} = \lambda.$$

We used change of variables $m = k - 1$.

Solutions for problems for Week 6

Solution of Problem 6.1. By the properties of the correlation and variance,

$$\begin{aligned}\mathrm{Var}\,(X+Y) &= \mathrm{Var}\,X + 2\mathrm{corr}(X,Y)\sigma_X\sigma_Y + \mathrm{Var}\,Y\\ &= 4 - 2\cdot 0.5\cdot 2\cdot 3 + 9 = 4 - 6 + 9 = 7.\end{aligned}$$

Solution of Problem 6.2. By the properties of the correlation and variance,

$$\begin{aligned}\mathrm{Var}\,(X-Y) &= \mathrm{Var}\,X + 2\mathrm{corr}(X,-Y)\sigma_X\sigma_Y + \mathrm{Var}\,Y\\ &= 4 + 2\cdot 0.5\cdot 2\cdot 3 + 9 = 4 + 6 + 9 = 19.\end{aligned}$$

Solution of Problem 6.3. By the properties of the correlation and variance,

$$\mathrm{corr}(X,-Y) = -\mathrm{corr}(X,Y),$$

and

$$\mathrm{Var}\,(X-Y-1) = \mathrm{Var}\,(X-Y).$$

Hence

$$\begin{aligned}\mathrm{Var}\,(X-Y-1) = \mathrm{Var}\,(X-Y) &= \mathrm{Var}\,X - 2\mathrm{corr}(X,Y)\sigma_X\sigma_Y + \mathrm{Var}\,Y\\ &= 25 - 2\cdot 0.5\cdot 5\cdot 3 + 9 = 25 - 15 + 9 = 19.\end{aligned}$$

Solution for Problem 6.4. We have that

$$\begin{aligned}\mathbf{E}Y = 0, \quad \mathrm{Var}\,Y = 1, \quad \mathrm{Cov}(X,Y) = 0.2\cdot 1\cdot\sqrt{2},\\ \mathbf{E}(XY) = \mathrm{Cov}(X,Y) + \mathbf{E}X\mathbf{E}Y = 0.2\sqrt{2}.\end{aligned}$$

Solution of problem 6.5. Since $Y = \mathbb{I}_{\{|X|<1/2\}}$, this random variables depends on each other, i.e., they are not independent. We have that $\mathbf{E}X = 0$, hence $\mathrm{Cov}(X,Y) = \mathbf{E}(XY)$. Since $Y = \mathbb{I}_{\{|X|<1/2\}}$, this random variables are not independent. However, we obtain that

$$\begin{aligned}&\mathrm{Cov}(X,Y) = \mathbf{E}(XY)\\ &= \int_{-1}^{1} x\mathbb{I}_{\{|x|<1/2\}} f_X(x)dx = \frac{1}{2}\int_{-1/2}^{1/2} xdx = \frac{x^2}{4}\bigg|_{-1/2}^{1/2} = 0.\end{aligned}$$

Solution for Problem 6.6. By the properties of the correlation and variance,

$$\text{corr}(X, -Y) = \text{corr}(X, Y),$$

and

$$\text{Var}\,(X - Y - 1) = \text{Var}\,(X - Y).$$

Hence

$$\begin{aligned}\text{Var}\,(X - Y - 1) = \text{Var}\,(X - Y) &= \text{Var}\,X - 2\text{corr}(X, Y)\sigma_X\sigma_Y + \text{Var}\,Y \\ &= 25 - 2 \cdot 0.5 \cdot 5 \cdot 3 + 9 = 25 - 15 + 9 = 19.\end{aligned}$$

Solution for Problem 6.7. We found in Example 3.6 that $\text{mes}\,(D) = 1/2$ and that the joint density is

$$f(x, y) = (1/2)^{-1}\mathbb{I}_D(x, y) = 2\mathbb{I}_D(x, y).$$

Also, we found that

$$f_X(x) = \int_{-\infty}^{\infty} f(x, y)dy = \int_0^x 2dy = 2x, \quad x \in [0, 1].$$

$$f_Y(y) = \int_{-\infty}^{\infty} f(x, y)dx = \int_y^1 2dx = 2(1 - y), \quad y \in [0, 1].$$

Hence

$$\mathbf{E}X = \int_{-\infty}^{\infty} x f_X(x)dx = \int_0^1 x2xdx = 2/3,$$

$$\begin{aligned}\mathbf{E}Y = \int_{-\infty}^{\infty} y f_Y(y)dy = \int_0^1 y2(1 - y)dy &= 2\int_0^1 ydy - 2\int_0^1 y^2dy \\ &= 1 - 2/3 = 1/3,\end{aligned}$$

$$\mathbf{E}X^2 = \int_{-\infty}^{\infty} x^2 f_X(x)dx = \int_0^1 x^2 2xdx = 2/4 = 0.5,$$

$$\begin{aligned}\mathbf{E}Y^2 = \int_{-\infty}^{\infty} x f_Y(y)dy = \int_0^1 y^2 2(1 - y)dy &= 2\int_0^1 y^2dy - 2\int_0^1 y^3dy \\ &= 2/3 - 2/4 = 0.1667.\end{aligned}$$

Hence $\mathrm{Var}\, X = 0.5-(2/3)^2 = 0.0556$, $\mathrm{Var}\, Y = 0.1667-(1/3)^2 = 0.0556$, Further,

$$\begin{aligned}\mathrm{Cov}(X,Y) &= \mathbf{E}(X-\mathbf{E}X)(Y-\mathbf{E}Y) = \int_D (x-2/3)(y-1/3)f(x,y)dx \\ &= 2\int_0^1 dx \int_0^x (x-2/3)(y-1/3)dy = 2\int_0^1 dx(x-2/3)\frac{(y-1/3)^2}{2}\Big|_0^x \\ &= 2\int_0^1 (x-2/3)\left(\frac{(x-1/3)^2}{2} - \frac{(1/3)^2}{2}\right)dx \\ &= \int_0^1 (x-2/3)\left((x-1/3)^2 - 1/9\right)dx = 1/36.\end{aligned}$$

Alternatively, one may use the formula

$$\mathrm{Cov}(X,Y) = \mathbf{E}(XY) - \mathbf{E}X\mathbf{E}Y.$$

We have

$$\mathbf{E}(XY) = \int_D xyf(x,y)dx = 2\int_0^1 dx\int_0^x xydy = 2\int_0^1 dxx\frac{x^2}{2} = \frac{1}{4}.$$

It gives $\mathrm{Cov}(X,Y) = \mathbf{E}(XY) - \mathbf{E}X\mathbf{E}Y = 1/4 - (2/3)(2/3) = 1/4 - 2/9 = 9/36 - 8/36 = 1/36$ again.

Solutions for problems for Week 7

Solution for Problem 7.1. We have that

$$\begin{aligned}\mathbf{E}(X|Y) &= \mathbf{E}(a + bY + cY^2 + \varepsilon|Y) = \mathbf{E}(a + bY + cY^2|Y) + b + \mathbf{E}(\varepsilon|Y) \\ &= a + bY + cY^2 + \mathbf{E}\varepsilon = a + bY + cY^2.\end{aligned}$$

We use here that $\mathbf{E}(\varepsilon|Y) = \mathbf{E}\varepsilon = 0$, by independence of Y and ε.

Solution for Problem 7.2. We have that

$$X = (Y+\varepsilon)^3.$$

Hence

$$\mathbf{E}(X|Y=y) = \mathbf{E}((Y+\varepsilon)^3|Y=y) = \mathbf{E}((y+\varepsilon)^3|Y=y) = \mathbf{E}(y+\varepsilon)^3.$$

We use here the independence of Y and ε.

$$\begin{aligned}\mathbf{E}(X|Y=y) &= \int_0^1 (y+u)^3du = \int_0^1 (y^3 + 3y^2u + 3yu^2 + u^3)du \\ &= y^3 + 3y^2/2 + y + 1/4.\end{aligned}$$

Hence

$$\mathbf{E}(X|Y) = Y^3 + 3Y^2/2 + Y + 1/4.$$

Solution of Problem 7.3. We have that

$$\eta(\omega) = -0.1, \quad \omega = \omega_2,$$
$$\eta(\omega) = 0.1, \quad \omega = \omega_1 \quad \text{or} \quad \omega = \omega_3.$$

We have that $\mathbf{E}\{\xi \,|\, \eta\} = \hat{\zeta}$, where $\hat{\zeta} = \hat{h}(\eta)$ for some function $\hat{h} : \mathbf{R} \to \mathbf{R}$ such that $\mathbf{E}(\zeta - \xi)^2$ is minimal over functions $h : \mathbf{R} \to \mathbf{R}$. Any function $\zeta = h(\eta)$ has a form $\zeta(\omega) = \alpha$ for $\omega = \omega_2$ and $\zeta(\omega) = \beta$ for $\omega = \omega_1$ or $\omega = \omega_3$, $\alpha, \beta \in \mathbf{R}$. Let $f(\alpha, \beta) = \mathbf{E}(\zeta - \xi)^2$. It suffices to find (α, β) such that $f(\alpha, \beta) = \mathbf{E}(\zeta - \xi)^2 = \min$. We have that

$$\begin{aligned} f(\alpha, \beta) &= \mathbf{E}(\zeta - \xi)^2 \\ &= \mathbf{P}(\{\omega_1\})(\beta - 0.1)^2 + \mathbf{P}(\{\omega_2\})(\alpha - 0)^2 + \mathbf{P}(\{\omega_3\})(\beta + 0.1)^2 \\ &= \frac{1}{4}(\beta - 0.1)^2 + \frac{1}{4}\alpha^2 + \frac{1}{2}(\beta + 0.1)^2. \end{aligned}$$

We have that

$$\frac{\partial f}{\partial \alpha}(\alpha, \beta) = \frac{\alpha}{2}, \quad \frac{\partial f}{\partial \beta}(\alpha, \beta) = \frac{1}{2}(\beta - 0.1) + \beta + 0.1 = \frac{3}{2}\beta + 0.05.$$

Hence the only minimum of f is at $(\alpha, \beta) = (0, -0.1/3)$. Therefore,

$$\mathbf{E}\{\xi \,|\, \eta\}(\omega) = 0, \quad \omega = \omega_2,$$
$$\mathbf{E}\{\xi \,|\, \eta\}(\omega) = 1/3, \quad \omega = \omega_1 \quad \text{or} \quad \omega = \omega_3.$$

Clearly, $\mathbf{E}\{\xi \,|\, \eta\}$ can be expressed as a deterministic function of η:

$$\mathbf{E}\{\xi \,|\, \eta\} = h(\eta), \quad \text{where} \quad h(-0.1) = 0, \quad h(x) = 1/30, \quad x \neq -0.1.$$

Solutions for problems for Week 8

Solution for Problem 8.1. We have that $X = e^{\log 2Y}$. Hence $\mathbf{E}X = M_Y(\log 2)$, where $M_Y(t)$ is moment generating function for Y. The moment generating function is

$$\begin{aligned} M_Y(t) = \mathbf{E}e^{tY} &= \alpha \int_0^\infty e^{tx} e^{-\alpha x} dx = \alpha \int_0^\infty e^{(t-\alpha)x} dx \\ &= \frac{\alpha}{t - \alpha} e^{(t-\alpha)x} \bigg|_0^\infty = -\frac{\alpha}{t - \alpha} = \frac{3}{3 - t}. \end{aligned}$$

It is defined for $t-6<0$. (Alternatively, we could use reference tables.) It gives

$$\mathbf{E}X = \frac{3}{3-\log 2} = \frac{3}{3-0.6931} = 1.3005.$$

Solution for Problem 8.2. We have that the m.g.f. is

$$M_X(t) = \frac{e^t+1}{2-t} = M_1(t)M_2(t),$$

where $M_1(t) = 0.5e^t+0.5 = pe^t+1-p$, $p=0.5$, $M_2(t) = 2/(2-t)$. We have that $M_1(t)$ is moment generating function of a Bernoulli random variable Y such that

$$Y = \begin{cases} 1, & \text{with probability } 0.5 \\ 0, & \text{with probability } 0.5. \end{cases}$$

We have that $M_2(t)$ is moment generating function of a random variable Z distributed as Exp(2). Hence X has the same distribution as $Z+Y$, where Z and Y are independent. Hence $\mathbf{E}X = \mathbf{E}Y+\mathbf{E}Z = 0.5+0.5 = 1$. Further,

$$\begin{aligned} \mathbf{E}X^2 &= \mathbf{E}(Y+Z)^2 = \mathbf{E}Y^2 + 2\mathbf{E}(YZ) + \mathbf{E}(Z^2) \\ &= \mathbf{E}(Y+Z)^2 = \mathbf{E}Y^2 + 2\mathbf{E}Y\cdot\mathbf{E}Z + \mathbf{E}(Z^2) \\ &= 0.5 + 2\cdot 0.5\cdot 0.5 + 0.5 = 1.5. \end{aligned}$$

Solution for Problem 8.3. We have that the m.g.f. is

$$M_X(t) = \frac{e^{3t}+e^{4t}}{2-t} = M_1(t)M_2(t)M_3(t),$$

where $M_1(t) = 0.5e^t+0.5 = pe^t+1-p$, $p=0.5$, $M_2(t) = 2/(2-t)$, $M_3(t) = e^{3t}$. We have that $M_1(t)$ is the moment generating function of a Bernoulli random variable Y such that

$$Y = \begin{cases} 1, & \text{with probability } 0.5 \\ 0, & \text{with probability } 0.5. \end{cases}$$

We have that $M_2(t)$ is the moment generating function of a random variable Z distributed as Exp(2). We have that $M_3(t)$ is the moment generating function of a non-random constant $C=3$. Hence X has the same distribution as $Z+Y+C$, where Z and Y are independent. Hence $\mathbf{E}X = \mathbf{E}Y+\mathbf{E}Z+C = 0.5+0.5+3 = 4$.

Solution for Problem 8.4. (a) The c.d.f. for X_i is $F(x) = (1 - e^{-2x})\mathbb{I}_{[0,+\infty)}(x)$. It can be calculated as $\mathbf{P}(X_i > x) = \int_x^\infty \alpha e^{-2s} ds = e^{-2x}$. The p.d.f. for Y is

$$f_Y(y) = 3(1 - F(y))^2 f(y) = 3e^{-4y} 2e^{-2y} \mathbb{I}_{[0,+\infty)}(y) = 6\lambda e^{-6y} \mathbb{I}_{[0,+\infty)}(y).$$

It is an exponential distribution again; the c.d.f. is

$$F_Y(y) = (1 - e^{-\alpha y})\mathbb{I}_{[0,+\infty)}(y).$$

Let $\alpha = 6$. The moment generating function is

$$M(s) = \mathbf{E}e^{sY} = \alpha \int_0^\infty e^{sx} e^{-\alpha x} dx = \alpha \int_0^\infty e^{(s-\alpha)x} dx$$

$$= \frac{\alpha}{s - \alpha} e^{(s-\alpha)x} \Big|_0^\infty = -\frac{\alpha}{s - \alpha} = \frac{6}{6 - s}.$$

It is defined for $s - 6 < 0$.

(b) $\mathbf{E}(Y)$ and $\mathrm{Var}\,(Y)$ can be found from the tables. However, we will calculate them using the moment generating functions:

$$\mathbf{E}Y = M'(0) = \frac{\alpha}{(\alpha - s)^2} \Big|_{s=0} = \frac{1}{\alpha} = \frac{1}{6} = 0.1667$$

and

$$\mathbf{E}Y^2 = M''(0) = \frac{2\alpha}{(\alpha - s)^3} \Big|_{s=0} = \frac{2}{\alpha^2} = \frac{1}{18} = 0.05556.$$

Solution for Problem 8.5. (a)

$$M(t) = \mathbf{E}e^{tX} = \int_1^\infty c e^{tx} e^{-2x} dx = \frac{ce^{(t-2)x}}{t - 2} \Big|_{t=1}^\infty = \frac{ce^{(t-2)x}}{t - 2} \Big|_{x=+\infty} - \frac{ce^{(t-2)}}{t - 2}$$

$$= \frac{ce^{(t-2)}}{2 - t}.$$

It gives

$$M(0) = \frac{ce^{-2}}{2}.$$

Since $M(0) = 1$, it implies that $c = 2e^2$ and

$$M(t) = 2\frac{e^t}{2 - t}.$$

(b)

$$M'(t) = 2\frac{e^t(2 - t) + e^t}{(2 - t)^2}, \quad M'(0) = 3/2 = 1.5.$$

(c) We have that

$$M(t) = e^t \frac{2}{2-t} = e^{1 \cdot t} M_Y(t),$$

where

$$M_Y(t) = \frac{2}{2-t}$$

is the m.g.f. of the exponential random variable Y with the density

$$f_Y(x) = \begin{cases} 2e^{-2x}, & x \geq 0 \\ 0, & \text{otherwise.} \end{cases}$$

Hence $X = 1 + Y$. (It can be noted that we could solve (b) easier using this.)

Solutions for problems for Week 9

Solution for Problem 9.1. (a) We have $\sigma_X^2 = 3$, $\sigma_Y^2 = 2$, $\mu_X = 7$, $\mu_Y = 5$. By the theorem from Week 9, we have

$$\mathbf{E}(X|Y) = 7 + (-0.2)\frac{\sqrt{3}}{\sqrt{2}}(Y - 5).$$

(b) Assume that we observe that $Y = 0.3$. Then the best estimate of X is

$$\mathbf{E}(X|Y = 0.3) = 7 + (-0.2)\frac{\sqrt{3}}{\sqrt{2}}(0.3 - 5) = 8.15.$$

(c) $Z \sim N(\mu_z, \sigma_Z^2)$, where

$$\begin{aligned}
\mu_Z &= 2 + 0.5\mu_X + 0.1\mu_Y = 6, \\
\sigma_Z^2 &= 0.25\sigma_X^2 + 2 \cdot 0.5 \cdot 0.1\text{corr}(X, Y)\sigma_X \sigma_y + 0.01\sigma_Y^2 \\
&= 0.25 \cdot 3 - 2 \cdot 0.2 \cdot 0.5 \cdot 0.1\sqrt{3}\sqrt{2} + 0.01 \cdot 2 = 0.721, \\
\sigma_Z &= \sqrt{0.721} = 0.8491.
\end{aligned}$$

Hence $Z = 6 + 0.8491U$, where $U \sim N(0, 1)$. Hence

$$\begin{aligned}
\mathbf{P}(Z < 5) &= \mathbf{P}(6 + 0.8491U < 5) = \mathbf{P}(U < -1/0.8491) = \mathbf{P}(U < -1.178) \\
&= \mathbf{P}(U > 1.178) = 1 - \mathbf{P}(U < 1.178) = 1 - 0.8810 = 0.119.
\end{aligned}$$

(Table A2; in this table, there are no negative numbers such as –1.178). Some other tables allow to calculate $\mathbf{P}(U < -1.178)$ directly.

Solution for Problem 9.2. We have that $U \sim \chi_2^2$. We obtain from the tables that $c = 4.65$:

degrees of freedom (v)	..	0.1	...	....
..	...	...	....	...
2	...	4.65	...	...

Solution for Problem 9.3. (a) $U \sim \chi_3^2$. We find from the table

degrees of freedom (v)	..	0.1	...	....
..	...	...	....	...
3	...	6.251	...	...

It gives $\mathbf{P}(W > 6.25) = 0.1$ and $\mathbf{P}(W < 6.25) = 0.9$.

(b) We have $W/2 \sim \chi_3^2$. We have that

$$\mathbf{P}(W < 0.432) = \mathbf{P}(W/2 < 0.216).$$

We find from the table

degrees of freedom (v)	..	0.975	...	....
..	...	...	....	...
3	...	0.216	...	...

It gives $\mathbf{P}(W > 0.432) = 0.975$ and $\mathbf{P}(W < 0.432) = 0.025$.

(c) We have that $X_i \sim 0.5Y_i$, where $Y_i \sim N(0,1)$. Hence

$$V = 0.25(Y_1^2 + Y_2^2 + Y_3^2).$$

Hence $4V \sim \chi_3^2$. We have that

$$\mathbf{P}(V > 1.95) = \mathbf{P}(4V > 7.8).$$

We find from table A3

degrees of freedom (v)	..	0.05	...	....
..	...	...	....	...
3	...	7.815	...	...

It gives $\mathbf{P}(V < 1.95) = 0.95$.

Solution for Problem 9.4. We have that $4V \sim \chi_3^2 \sim \Gamma(3/2, 1/2)$. Hence

$$\mathbf{E}((4V)^{1/2}) = \int_0^\infty \frac{\lambda^\alpha}{\Gamma(\alpha)} x^{1/2} x^{\alpha-1} e^{-\lambda x} dx$$

for $\alpha = 3/2$, $\lambda = 1/2$, or

$$\mathbf{E}((4V)^{1/2}) = \frac{\lambda^\alpha}{\Gamma(\alpha)} \int_0^\infty x^{1/2} x^{3/2-1} e^{-x/2} dx = \frac{\lambda^\alpha}{\Gamma(\alpha)} \int_0^\infty x e^{-x/2} dx$$

$$= \frac{\lambda^\alpha}{\Gamma(\alpha)} \frac{\Gamma(2)}{\lambda^2} \frac{\lambda^2}{\Gamma(2)} \int_0^\infty x e^{-x/2} dx = \frac{\lambda^\alpha}{\Gamma(\alpha)} \frac{\Gamma(2)}{\lambda^2} = \frac{0.5^{3/2}}{0.886227} \frac{1}{1/4} = 1.5958.$$

For the values of the Gamma function, we can MATLAB or other software, or any Gamma function calculator available online. We obtained above that $2\mathbf{E}((V)^{1/2}) = 1.5958$. This gives $\mathbf{E}((V)^{1/2}) = 1.5958/2 = 0.7979$.

Solutions for problems for Week 10

Solution for Problem 10.1. (a) We have that $\mathbf{E}X_k = \int_0^1 2x^2 dx = \frac{2}{3}$,

$$\mathbf{E}Z_n = n^{-1} \sum_{i=1}^n \mathbf{E}X_k = n^{-1} n \frac{2}{3} = \frac{2}{3},$$

$$\mathbf{E}X_k^2 = \int_0^1 2x^3 dx = \frac{1}{2}, \qquad \operatorname{Var} X_k = 1/2 - (2/3)^2 = 1/18,$$

and

$$\operatorname{Var} Z_n = n^{-1} \sum_{i=1}^n \mathbf{E} \operatorname{Var} X_k = n^{-2} n \frac{1}{18} = \frac{1}{18n}.$$

(b) The Law of Large Numbers gives that $Z_n \to 2/3$ as $n \to +\infty$ in probability, i.e., $\mathbf{P}(|Z_n - 2/3| > \varepsilon) \to 0$ for any $\varepsilon > 0$. The Central Limit Theorem gives that

$$\mathbf{P}\left(\frac{Z_n - 2/3}{1/\sqrt{18n}} \le x\right) \to \mathbf{P}(Y \le x),$$

where $Y \sim N(0,1)$, for all x.

(c) Let us estimate $\mathbf{P}(Z_{16} > 0.7)$. By the Central Limit Theorem, we obtain that

$$\mathbf{P}\left(\frac{Z_n - 2/3}{1/\sqrt{18n}} > x\right) \to \mathbf{P}(Y > x) \quad \text{as} \quad n \to +\infty,$$

where $Y \sim N(0,1)$. Clearly,

$$\mathbf{P}(Z_n > 0.7) = \mathbf{P}\left(\frac{Z_n - 2/3}{1/\sqrt{18n}} > \frac{0.7 - 2/3}{1/\sqrt{18n}}\right).$$

For $n = 16$, the Central Limit Theorem gives that

$$\mathbf{P}(Z_{16} > 0.7) \sim \mathbf{P}\left(Y > \frac{1/30}{1/(4\sqrt{18})}\right).$$

Using the tables for the standard normal c.d.f., we obtain

$$\mathbf{P}(Z_{16} > 0.7) \sim \mathbf{P}\left(Y > \frac{1/30}{1/(4\sqrt{18})}\right) = \mathbf{P}(Y > 0.5657) = 1 - 0.7157$$
$$= 0.2843.$$

Solution for Problem 10.2 (a) It is an exponential distribution with the parameter $\lambda = 5$. It follows that $\mathbf{E}X_k = \frac{1}{5}$, $\mathrm{Var}\, X_k = 1/25$, and

$$\mathrm{Var}\, Z_n = n^{-1} \sum_{i=1}^{n} \mathbf{E}\mathrm{Var}\, X_k = n^{-2} n \frac{1}{25} = \frac{1}{25n}.$$

(b) Law of Large Numbers is applicable: $Z_n \to 1/5$ as $n \to +\infty$ in probability, i.e., $\mathbf{P}(|Z_n - 1/5| > \varepsilon) \to 0$ for any $\varepsilon > 0$. The Central Limit Theorem gives that

$$\mathbf{P}\left(\frac{Z_n - 1/5}{1/\sqrt{25n}} \le x\right) \to \mathbf{P}(Y \le x),$$

where $Y \sim N(0,1)$, for all x.

(c) Let us estimate $\mathbf{P}(Z_{11} > 0.25)$. Clearly,

$$\mathbf{P}(Z_n > 0.25) = \mathbf{P}\left(\frac{Z_n - 1/5}{1/\sqrt{25n}} > \frac{0.25 - 0.2}{1/\sqrt{25n}}\right).$$

For $n = 11$, the Central Limit Theorem gives

$$\mathbf{P}(Z_{11} > 0.25) \sim \mathbf{P}\left(Y > \frac{0.25 - 0.2}{1/\sqrt{25 \cdot 11}}\right) = \mathbf{P}(Y > 0.8292).$$

Using the tables for the standard normal c.d.f., we obtain

$$\mathbf{P}(Z_{11} > 0.25) \sim \mathbf{P}(Y > 0.8292) = 1 - 0.7967 = 0.2033.$$

Solution for Problem 10.3.

(a) Let X be the gain. It is a random variable that takes value 35 if number 1 comes up, value -1 otherwise. Hence

$$\begin{aligned}
\mathbf{P}(X = -1) &= 37/38, \quad \mathbf{P}(X = 35) = 1/38, \\
\mu = \mathbf{E}X &= (-1)37/38 + 35/38 = -1/19, \\
\mathbf{E}(X^2) &= 1 \cdot 37/38 + 35^2 \cdot 1/38 = 33.211, \\
\sigma^2 = \mathrm{Var}\, X &= \mathbf{E}(X^2) - (\mathbf{E}X)^2 = 33.208.
\end{aligned}$$

(b) Let

$$Z_n = n^{-1} \sum_{k=1}^{n} X_k,$$

where X_k are independent and distributed as X in (a).

Let us estimate $\mathbf{P}(Z_{20} > 0)$. Clearly,

$$\mathbf{P}(Z_n > 0) = \mathbf{P}\left(\frac{Z_n - \mu}{\sigma/\sqrt{n}} > \frac{0 - (-1/19)}{\sigma/\sqrt{n}}\right).$$

For $n = 20$, Central Limit Theorem gives

$$\mathbf{P}(Z_{20} > 0) \sim \mathbf{P}\left(Y > \frac{1/19}{\frac{\sqrt{32.208}}{\sqrt{20}}}\right) = \mathbf{P}(Y > 0.0408).$$

Using the tables for the standard normal c.d.f., we obtain

$$\mathbf{P}(Z_{20} > 0) \sim \mathbf{P}(Y > 0.0408) = 1 - \mathbf{P}(Y < 0.0408) = 1 - 0.5160 = 0.4840.$$

Solution for Problem 10.4 (a) Let X be the gain. It is a random variable that takes value 1 if red comes up, value –1 otherwise. Hence

$$\begin{aligned}
&\mathbf{P}(X = -1) = 20/38, \quad \mathbf{P}(X = 1) = 18/38,\\
&\mu = \mathbf{E}X = (-1)20/38 + 18/38 = -1/19,\\
&\mathbf{E}(X^2) = 1 \cdot 20/38 + 1 \cdot 18/38 = 1,\\
&\sigma^2 = \operatorname{Var} X = \mathbf{E}(X^2) - (\mathbf{E}X)^2 = 1 - (1/19)^2 = 0.997.
\end{aligned}$$

(b) Let

$$Z_n = n^{-1} \sum_{k=1}^{n} X_k,$$

where X_k are independent and distributed as X in (a).

Let us estimate $\mathbf{P}(Z_{20} > 0)$. Clearly,

$$\mathbf{P}(Z_n > 0) = \mathbf{P}\left(\frac{Z_n - \mu}{\sigma/\sqrt{n}} > \frac{0 - (-1/19)}{\sigma/\sqrt{n}}\right).$$

For $n = 20$, the Central Limit Theorem gives

$$\mathbf{P}(Z_{20} > 0) \sim \mathbf{P}\left(Y > \frac{1/19}{\sqrt{0.997}/\sqrt{20}}\right) = \mathbf{P}(Y > 0.2357).$$

Using the tables for the standard normal c.d.f., we obtain

$$\mathbf{P}(Z_{20} > 0) \sim \mathbf{P}(Y > 0.23) = 1 - \mathbf{P}(Y < 0.23) = 1 - 0.5910 = 0.4090.$$

Note that for the game from Problem 10.3 we obtained larger probability $\mathbf{P}(Z_{20} > 0) \sim 0.4840$. It can be explained as follows: that $\mathbf{P}(Z_n > 0) \to 0$ as $n \to +\infty$ in both cases but the rate of convergence is less for the case of higher variance (i.e., with betting on a single number).

Solutions for problems for Week 11

Solution for Problem 11.1. (a) Let $n = 3$ and $(X_1, X_2, X_3) = (-1, -1, 1)$. Then

$$\overline{X} = \frac{1}{3}(-1 - 1 + 1) = -\frac{1}{3},$$
$$S^2 = \frac{1}{3-1}\left(\left(-1 + \frac{1}{3}\right)^2 + \left(-1 + \frac{1}{3}\right)^2 + \left(1 + \frac{1}{3}\right)^2\right) = \frac{4}{3}.$$

(b) $\overline{X} \sim N(2, 3/n$, $n = 3$, i.e., $\overline{X} \sim N(2, 1)$. We have $Y = X - 2 \sim N(0, 1)$. We obtain

$$\mathbf{P}(\overline{X} < 2.2) = \mathbf{P}(Y < 0.2) = 0.5793.$$

(c) $\frac{2}{3}S^2 \sim \chi_2^2$. We have

$$\mathbf{P}\left(S^2 < 6.9\right) = \mathbf{P}\left(\frac{2}{3}S^2 < \frac{2}{3}6.51\right) = \mathbf{P}\left(U < 4.6\right),$$

where $U \sim \chi_2^2$ (with 2 degrees of freedom).

Let us use table A3

degrees of freedom (v)	..	$p = 0.1$	....	....
..	...	...	....	...
2	...	4.605	...	...

We obtain

$$\mathbf{P}\left(S^2 < 6.1\right) = \mathbf{P}\left(U < 4.6\right) \sim \mathbf{P}(U \leq 4.605) = 1 - \mathbf{P}(U > 4.605)$$
$$= 1 - 0.1 = 0.9.$$

Solution for Problem 11.2 (a) Let $n = 6$ and $(X_1, X_2, X_3, X_4, X_6) = (-1, -1, 0, 2, 1, 2)$. Then

$$\overline{X} = \frac{1}{6}(-1 - 1 + 0 + 2 + 1 + 2) = \frac{3}{6} = \frac{1}{2},$$

and

$$S^2 = \frac{1}{6-1}\left(\left(-1 - \frac{1}{2}\right)^2 + \left(-1 - \frac{1}{2}\right)^2 + \left(0 - \frac{1}{2}\right)^2 + \left(2 - \frac{1}{2}\right)^2 + \left(1 - \frac{1}{2}\right)^2 + \left(2 - \frac{1}{2}\right)^2\right) = 1.9.$$

(b) $\overline{X} \sim N(2, 4/n, n = 6$, i.e., $\overline{X} \sim N(2, 2/3)$. We have $X = 2 + Y \cdot \sqrt{2/3}$, i.e. $Y = (X-2)/\sqrt{2/3}$, where $Y \sim N(0,1)$. We obtain

$$\mathbf{P}(\overline{X} < 2.2) = \mathbf{P}\left(Y < \frac{2.2-2}{\sqrt{2/3}}\right) = \mathbf{P}(Y < 0.2449) = 0.5967.$$

(c) $\frac{5}{4}S^2 \sim \chi_5^2$. We have

$$\mathbf{P}\left(S^2 < 0.484\right) = \mathbf{P}\left(\frac{5}{4}S^2 < \frac{5}{4}0.484\right) = \mathbf{P}\left(U < 0.605\right),$$

where $U \sim \chi_5^2$ (with 5 degrees of freedom).

Let us use table A3

degrees of freedom (v)	..	$p = 0.99$	....	....
..	...	...	....	...
5	...	0.554	...	...

We obtain

$$\mathbf{P}\left(S^2 < 0.484\right) = \mathbf{P}\left(U < 0.605\right) \sim \mathbf{P}(U \le 0.554) = 1 - 0.99 = 0.01.$$

Solution for Problem 11.3. (a) $\overline{X} \sim N(2, 3/n)$, $n = 6$, i.e., $\overline{X} \sim N(2, 1/2)$. We have $X = 2 + Y/2$, i.e., $Y = 2(X-2)$, where $Y \sim N(0,1)$. We obtain

$$\mathbf{P}(\overline{X} < 2.2) = \mathbf{P}(Y < 2(2.2 - 0.2)) = \mathbf{P}(Y < 0.4) = 0.6554.$$

(b) $\frac{5}{3}S^2 \sim \chi_5^2$. We have

$$\mathbf{P}\left(S^2 < 6\right) = \mathbf{P}\left(\frac{5}{3}S^2 < \frac{5}{3}6\right) = \mathbf{P}\left(U < 10\right),$$

where $U \sim \chi_5^2$ (with 5 degrees of freedom).

Let us use table A3

degrees of freedom (v)	..	$p = 0.1$	....	....
..	...	...	....	...
5	...	9.236	...	...

It gives

$$\mathbf{P}\left(S^2 < 6\right) = \mathbf{P}\left(U < 10\right) \sim \mathbf{P}(U \le 9.236) = 1 - 0.1 = 0.9.$$

Solutions for problems for Week 12

Solution of Problem 12.1. Let $\hat{p} = 0.5$ be the proportion obtained for the sample of size n, with $n = 250$, and let p be the true unknown proportion. We have that

$$n\hat{p} \sim Bin(n,p), \quad \mathbf{E}\hat{p} = p, \quad \mathrm{Var}\,\hat{p} = p(1-p)/n.$$

The distribution of $\hat{p}$ can be considered approximately as $N(p, p(1-p)/n)$. Since p is unknown, we have to approximate it again by $N(p, \hat{p}(1-\hat{p})/n)$. Therefore, we accept that

$$X = \frac{\hat{p}-p}{\sqrt{\hat{p}(1-\hat{p})n}}$$

is distributed as $N(0,1)$, and that, for $\alpha \in (0,1)$, we have

$$\mathbf{P}\left(-z_{\alpha/2} < X < z_{\alpha/2}\right) = 1-\alpha.$$

For this example, we have $\alpha = 0.1$, and

$$z_{\alpha/2} = 1.65, \quad \sqrt{\hat{p}(1-\hat{p})/n} = \sqrt{0.25/250} = 0.0316,$$
$$z_{\alpha/2}\sqrt{\hat{p}(1-\hat{p})/n} = 0.0522.$$

Hence

$$\begin{aligned} 1-\alpha &= \mathbf{P}\left(\hat{p} - z_{\alpha/2}\sqrt{\hat{p}(1-\hat{p})/n} < p < \hat{p} + z_{\alpha/2}\sqrt{\hat{p}(1-\hat{p})/n}\right) \\ &= \mathbf{P}\left(\hat{p} - 0.0522 < p < \hat{p} + 0.0522\right) \\ &= \mathbf{P}\left(0.5 - 0.0522 < p < 0.5 + 0.0522\right) \\ &= \mathbf{P}\left(0.4478 < p < 0.5522\right). \end{aligned}$$

Therefore, the interval

$$[0.4478, 0.5522]$$

is the 90%-confidence interval for the true unknown p.

Solution of Problem 12.2. We have that $\frac{(n-1)}{\sigma^2}S^2 \sim \chi^2_8$. Let $\alpha = 0.1$. We have that $\chi^2_{\alpha/2} = 15.507$ and $\chi^2_{1-\alpha/2} = 2.733$.

$$\mathbf{P}\left(\chi^2_{1-\alpha/2} < \frac{(n-1)}{\sigma^2}S^2 < \chi^2_{\alpha/2}\right) = 1-\alpha.$$

Hence

$$\mathbf{P}\left(1/\chi^2_{\alpha/2} < \frac{\sigma^2}{(n-1)S^2} < 1/\chi^2_{1-\alpha/2}\right) = 1-\alpha,$$

$$\mathbf{P}\left(\frac{(n-1)S^2}{\chi^2_{\alpha/2}} < \sigma^2 < \frac{(n-1)S^2}{\chi^2_{1-\alpha/2}}\right) = 1-\alpha.$$

This gives

$$1-\alpha = \mathbf{P}\left(\frac{8S^2}{\chi^2_{\alpha/2}} < \sigma^2 < \frac{8S^2}{\chi^2_{1-\alpha/2}}\right) = \mathbf{P}\left(\frac{24}{\chi^2_{\alpha/2}} < \sigma^2 < \frac{24}{\chi^2_{1-\alpha/2}}\right)$$
$$= \mathbf{P}\left(\frac{24}{15.507} < \sigma^2 < \frac{24}{2.733}\right) = \mathbf{P}\left(1.5388 < \sigma^2 < 8.7816\right).$$

Therefore, the interval

$$\left[1.5388 < \sigma^2 < 8.7816\right]$$

is the 90%-confidence interval for the true unknown σ^2.

Solution of Problem 12.3. Let $\mathbf{P}_0$ be the probability under H_0. Let $\hat{p}$ be the sample proportion. We have that

$$p - \text{value} = \mathbf{P}_0(\hat{p} \geq 0.4) = \mathbf{P}\left(X \geq \frac{0.6-p}{\sqrt{p(1-p)/n}}\right),$$

where $X = \frac{\hat{p}-p}{\sqrt{p(1-p)/n}}$ can be assumed to be from $N(0,1)$ under $\mathbf{P}_0$. Hence

$$p - \text{value} = \mathbf{P}_0\left(X \geq \frac{0.6-0.5}{\sqrt{0.5(1-0.5)/100}}\right) = \mathbf{P}_0\left(X \geq 2\right) = 0.02 < 0.05.$$

Therefore, H_0 has to be rejected.

Solution of Problem 12.4. We have that

$$\frac{\sigma_2^2 S_1^2}{\sigma_1^2 S_2^2} \sim F_{9,9}.$$

Hence, under H_0,

$$\frac{S_1^2}{S_2^2} \sim F_{9,9}.$$

The test statistics is $F = S_1^2/S_2^2 = 4$. We reject H_0 if $F \notin (f_{0.95}, f_{0.05})$, where $f_{0.05} = q_{0.95} = 3.179$, $f_{0.95} = q_{0.05} = 1/3.179 = 0.3146$ are the corresponding critical values for $F_{9,9}$. The acceptance interval is

$$(L, R) = (f_{0.95}, f_{0.05}) = (0.3146, 3.179).$$

Since $F \notin (q_{0.05}, f_{0.05})$, we reject H_0 with level of confidence 90%.

Solution of Problem 12.5. We have that

$$\frac{\sigma_2^2 S_1^2}{\sigma_1^2 S_2^2} \sim F_{10,12}.$$

Hence, under H_0,

$$\frac{S_1^2}{S_2^2} \sim F_{10,12}.$$

The test statistics is $F = S_1^2/S_2^2 = 0.5$. We reject H_0 if $F \notin (f_{0.95}, f_{0.05})$, where $f_{0.05} = 2.753$, $f_{0.95} = 1/f_{0.5}^{12,10} = 1/2.913 = 0.34$ are the corresponding critical values for the distribution $F_{10,12}$, and where $f_{0.05}^{12,10} = 2.913$ is the corresponding critical value for the distribution $F_{12,10}$. The acceptance interval is

$$(L, R) = (f_{0.95}, f_{0.05}) = (0.34, 2.753).$$

Since $F \in (f_{0.95}, f_{0.05})$, we cannot reject H_0.

Solutions for Problem 12.6. By the assumptions, the rejection region is

$$t = \sum_{i=1}^{n} x_i < c$$

for some c. Let us find this c. It suffices to find c such that

$$\alpha = 0.05 = \mathbf{P}_0\left(\sum_{i=1}^{n} X_i < c\right) = \mathbf{P}\left(\sum_{i=1}^{n} X_i < c \,\middle|\, \mu = 2\right)$$

where $\mathbf{P}_0$ is the probability under H_0. We have $\mathbf{E}(X_i|\mu = 2) = 2$ and $\mathrm{Var}\,(X_i|\mu) = \sigma^2$. Let

$$Z = \frac{\sum_{i=1}^{n}(X_i - 2)}{\sigma\sqrt{n}} = \frac{\sum_{i=1}^{n} X_i - 2n}{\sigma\sqrt{n}}.$$

We have that, under H_0, $Z \sim N(0, 1)$. Therefore, the tables can be used. We have

$$\alpha = \mathbf{P}_0\left(\sum_{i=1}^{n} X_i < c\right) = \mathbf{P}_0\left(Z < \frac{c - 2n}{\sigma\sqrt{n}}\right) = \Phi\left(\frac{c - 2n}{\sigma\sqrt{n}}\right).$$

It gives $(c - 2n)/\sigma\sqrt{n} = z_\alpha$, i.e., the αth quantile of $N(0, 1)$. For $\alpha = 0.05$, we have $z_{0.05} = -1.645$. For $n = 20$, $\sigma = 1$, $n = 20$, it gives

$$c - 40 = -1.645\sqrt{20},$$

or

$$c = 40 - 1.645\sqrt{20} = 32.64.$$

Solution of Problem 12.7. It suffices to find $c(\mu)$ such that

$$\alpha = 0.05 = \mathbf{P}\left(\sum_{i=1}^{n} X_i < c(\mu) \,\middle|\, \mu\right).$$

In this case,

$$1 - \alpha = 0.95 = \mathbf{P}\left(\sum_{i=1}^{n} X_i \geq c(\mu) \,\middle|\, \mu\right),$$

and the acceptance region is

$$A(\mu) = \left\{X = (X_1, ..., X_n) : \quad \sum_{i=1}^{n} X_i \geq c(\mu)\right\}.$$

We have that $\mathbf{E}(X_i|\mu) = \mu$ and $\mathrm{Var}\,(X_i|\mu) = \sigma^2$. Let

$$Z = \frac{\sum_{i=1}^{n}(X_i - \mu)}{\sigma\sqrt{n}} = \frac{\sum_{i=1}^{n} X_i - n\mu}{\sigma\sqrt{n}}.$$

Under H_0, $Z \sim N(0,1)$. Therefore, the tables can be used. We have

$$\alpha = \mathbf{P}_0\left(\sum_{i=1}^{n} X_i < c(\mu)\right) = \mathbf{P}_0\left(Z < \frac{c - n\mu}{\sigma\sqrt{n}}\right) = \Phi\left(\frac{c - n\mu}{\sigma\sqrt{n}}\right).$$

It gives $(c - n\mu)/\sigma\sqrt{n} = z_\alpha$, i.e., the αth quantile of $N(0,1)$. For $\alpha = 0.05$, we have $z_{0.05} = -1.645$. For $n = 20$, $\sigma = 1$, $n = 20$, it gives

$$c - 40 = -1.645\sqrt{20},$$

or

$$c = 40 - 1.645\sqrt{20} = 32.64.$$

For $n = 20$, $\sigma = 1$, $n = 20$, it gives

$$c(\mu) - 20\mu = -1.645\sqrt{20},$$

or

$$c(\mu) = 20\mu - 1.645\sqrt{20} = 20\mu - 7.36.$$

By the theorem from Week 12, we obtain the $100(1-\alpha)\%$ confidence region $C(X)$ for μ for $\alpha = 0.05$:

$$\mathbf{P}(\mu \in C(X)) = 1 - \alpha$$

where

$$C(X) = \left\{X = (X_1, ..., X_{20}) : \sum_{i=1}^{20} X_i \geq 20\mu - 7.36\right\}$$

or

$$\mathbf{P}\left(\mu \leq \frac{1}{20}\left[\sum_{i=1}^{20} X_i + 7.36\right]\right) \geq 1 - \alpha.$$

Appendix 2: Sample Problems for Final Exams

In this appendix, we list problems that may require the material from all Weeks 1-12. The problems of this type can be used for final exams.

Problem F.1 *Consider a casino roulette game. In this game, players make bets on certain numbers. The probability that an odd number come up in one game is* $18/38$*. Find the probability that this event happens twice during four games.*

Problem F.2 *Let* X_1*,* X_2*, and* X_2 *be independent random variables with the same density*

$$f(x) = \begin{cases} \frac{1}{3}e^{-x/3}, & x \geq 0 \\ 0, & \textit{otherwise.} \end{cases}$$

Let $Y = \min(X_1, X_2, X_3)$*. Calculate* $\mathbf{E}(Y)$ *and* $\mathbf{P}(Y \geq 0.5)$*.*

Problem F.3 *Let* X *has uniform distribution on* $(0, 1)$*, and* Y *has the density*

$$f_Y(y) = \begin{cases} e^{-y}, & y \geq 0 \\ 0, & y < 0. \end{cases}$$

We assume that X *and* Y *are independent. Let*

$$U = X - Y, \quad V = 2X.$$

Find the joint density $f_{UV}(u, v)$ *for* (U, V)*. Calculate its value at the points* $(u, v) = (1, 3)$ *and* $(u, v) = (0.25, 1)$ *(in particular, take into account the possibility that the density is zero at these points).*

Problem F.4 *Let random variables X and Y have the joint density $f(x,y)$ such that, for some constant c,*

$$f(x,y) = \begin{cases} cy^2, & 0 \le x \le 1,\ 0 \le y \le x \\ 0, & \textit{otherwise.} \end{cases}$$

Calculate $\mathbf{P}(X < 1/2)$.

Problem F.5 *Let X_1, X_2, X_3, X_4 be independent random variables with normal distribution $N(0,1)$. Let*

$$\overline{X} = \frac{1}{4}(X_1 + X_2 + X_3 + X_4),$$
$$Y = (X_1 - \overline{X})^2 + (X_2 - \overline{X})^2 + (X_3 - \overline{X})^2 + (X_4 - \overline{X})^2.$$

Find the moment generating function $M_Y(t)$ for Y. In particular, indicate the set of t where the function $M_Y(t)$ is defined.

Problem F.6 *Let $\{X_k\}$ be a sequence of independent random variables with the same distribution. Let*

$$Z_n = n^{-1} \sum_{k=1}^{n} X_k, \quad n = 1, 2, \ldots.$$

Assume that $\mathbf{E}(Z_9) = 0.4$ and $\mathrm{Var}(Z_9) = 0.25$. Using the Central Limit Theorem, estimate $\mathbf{P}(Z_9 > 0.45)$. Use the tables if necessary.

Problem F.7 *Let X be a random variable with the density*

$$f_X(x) = \begin{cases} c2^{-x}, & x \ge 1 \\ 0, & x < 1. \end{cases}$$

Here $c > 0$ is a constant. Calculate $\mathbf{E}(X)$.

Problem F.8 *Let X be a random variable with the density*

$$f(x) = \begin{cases} c\int_{\max(0,x)}^{x+1} ye^{-y}dy, & x \ge -1 \\ 0, & x < -1, \end{cases}$$

where $c > 0$ is a constant. Calculate $\mathbf{E}(X)$. Hint: the problem can be solved by direct calculation. Alternatively, the solution can be simplified by using some facts from the theory.

Problem F.9 *Let random variables X and Y be independent. Assume that X has normal distribution $N(2,9)$ and that Y has normal distribution $N(4,7)$. Calculate $\mathbf{P}(X + Y < 6.4)$.*

Problem F.10 *Let X_1, X_2, and X_2 be independent random variables with the same density*

$$f(x) = \begin{cases} 2e^{-2x}, & x \geq 0 \\ 0, & \textit{otherwise.} \end{cases}$$

Let $Y = \min(X_1, X_2, X_3)$. Find the moment generating function $M_Y(t)$ for Y. In particular, indicate the set of t where the function $M_Y(t)$ is defined.

Problem F.11 *Let X_1, X_2, X_3, X_4 be independent random variables with normal distribution $N(0, 1)$. Let*

$$\overline{X} = \frac{1}{4}(X_1 + X_2 + X_3 + X_4),$$
$$S^2 = \frac{1}{3}\left[(X_1 - \overline{X})^2 + (X_2 - \overline{X})^2 + (X_3 - \overline{X})^2 + (X_4 - \overline{X})^2\right].$$

Calculate $\mathbf{P}(S^2 \leq 2.08)$ using the tables.

Problem F.12 *Let X and Z be random variables with densities*

$$f_X(x) = \begin{cases} e^{-x}, & x \geq 0 \\ 0, & x < 0, \end{cases} \qquad f_Z(z) = \begin{cases} 0.5e^{-0.5z}, & z \geq 0 \\ 0, & z < 0. \end{cases}$$

Assume that the correlation of X and Z is -0.5, i.e. $\mathrm{corr}(X, Z) = -0.5$. Calculate $\mathbf{E}(XZ)$.

Problem F.13 *The sample variance S^2 was calculated for the sample of 11 independent random variables with the same normal distribution $N(\mu, \sigma^2)$. Find the 90%-confidence interval for σ^2 given that $S^2 = 4$.*

Problem F.14 *Consider an i.i.d. sample $X_1, ..., X_n$ from a population distribution $N(\mu, \sigma^2)$. Assume the sample mean $\overline{X}$, $\sigma^2 = 1$, and n are known. Find the $(1 - \alpha)100\%$-confidence interval for the true unknown μ. Assume that $\alpha = 0.1$, $\overline{X} = 3$, $\sigma^2 = 1$, and $n = 9$.*

Solutions

Solution of Problem F.1. Let X be the number trials where an odd number comes up. We have X has binomial distribution B(4,18/38). Hence

$$\mathbf{P}(X=2) = \binom{4}{2}\left(\frac{18}{38}\right)^2\left(\frac{20}{38}\right)^2 = 6\left(\frac{18}{38}\right)^2\left(\frac{20}{38}\right)^2 = 0.3729.$$

Solution of Problem F.2. We have that $X_i \sim Exp(\lambda)$ with $\lambda = 1/3$. The c.d.f. for X_i is $F(x) = (1-e^{-\lambda x})\mathbb{I}_{[0,+\infty)}(x)$. It can be calculated as $\mathbf{P}(X_i > x) = \int_x^\infty \lambda e^{-\lambda s} ds = e^{-\lambda x}$. The p.d.f. for Y is

$$\begin{aligned} f_Y(y) = 3(1-F(y))^2 f(y) = 3e^{-2\lambda y}\lambda e^{-\lambda y}\mathbb{I}_{[0,+\infty)}(y) &= 3\lambda e^{-3\lambda y}\mathbb{I}_{[0,+\infty)}(y) \\ &= e^{-y}\mathbb{I}_{[0,+\infty)}(y). \end{aligned}$$

It is an exponential distribution again; $\mathbf{E}(Y) = 1$ and $\mathbf{P}(Y > 0.5) = e^{-0.5} = 0.6065$.

Solution of Problem F.3. We have $u = g_1(x,y) = x-y$, $v = g_2(x,y) = x$. It gives

$$\begin{aligned} X &= h_1(U,V) = v/2, \\ Y &= h_2(U,V) = x - u = v/2 - u. \end{aligned}$$

The Jacobian is

$$J(u,v) = \det\begin{pmatrix} 0 & 1/2 \\ -1 & 1/2 \end{pmatrix} = 0\cdot 1/2 + 1\cdot 1/2 = 1/2.$$

The joint density for $(X < Y)$ is $f_{XY}(x,y) = \mathbb{I}_{[0,1]\times[0,+\infty)}(x,y)$. The joint density for U and V is

$$\begin{aligned} f_{UV}(u,v) &= f_{XY}(h_1(u,v), h_2(u,v))|J(u,v)| \\ &= 0.5 f_{XY}(v/2, v/2-u)\mathbb{I}_{\{v/2\in[0,1], v/2-u\geq 0\}}(u,v) \\ &= 0.5 f_{XY}(v/2, v/2-u)\mathbb{I}_{\{v\in[0,2], v/2>u\}}(u,v) \\ &= 0.5 f_{XY}(v/2, v/2-u)\mathbb{I}_{\{v\in[0,2], v>2u\}}(u,v). \end{aligned}$$

It is easy to see that if $(u,v) = (1,3)$, then $v > 2$. It follows that $\mathbb{I}_{\{v\in[0,2],v>2u\}}(u,v) = 0$ for $(u,v) = (1,3)$. Hence $f_{UV}(1,3) = 0$. Further, $\mathbb{I}_{\{v\in[0,2],v>2u\}}(u,v) = 1$ for $(u,v) = (0.25,1)$. Hence

$$f_{UV}(0.25,1) = 0.5\exp(-(v/2-u)) = 0.5\exp(-0.25) = 0.3894.$$

Solution of Problem F.4. The joint density is $f(x,y) = cy^2\mathbb{I}_D(x,y)$, where $D = \{(x,y) : x \in [0,1], \quad y \in [0,x]\}$. The constant c can be found from the equation

$$\int_{\mathbf{R}^2} f(x,y)dxdy = \int_D f(x,y)dxdy = c\int_D y^2 dxdy = 1.$$

We have

$$c\int_D y^2 dxdy = c\int_0^1 dx \int_0^x y^2 dy = c\int_0^1 \frac{x^3}{3}dx = c\frac{x^4}{12}\Big|_0^1 = c/12 = 1.$$

Hence $c = 12$. Further,

$$f_X(x) = \int_{-\infty}^{\infty} f(x,y)dy = \int_0^x 12y^2 dy = 4x^3, \quad x \in [0,1],$$

$$f_X(x) = 0, \quad x \notin [0,1].$$

Hence $\mathbf{P}(X < 1/2) = \int_0^{1/2} 4x^3 dx = 4\frac{x^4}{4}|_0^{1/2} = 1/16$.

Solution of Problem F.5.

$$S^2 = \frac{1}{3}\left[(X_1 - \overline{X})^2 + (X_2 - \overline{X})^2 + (X_3 - \overline{X})^2 + (X_4 - \overline{X})^2\right]$$

is the sample variance. Hence the random variable $Y = \frac{3}{1}S^2$ has χ^2-distribution with 3 degrees of freedom and

$$M_Y(t) = (1-2t)^{-3/2},$$

where $U \sim \chi_3^2$ (with degree of freedom 3).

Solution of Problem F.6. We accept that the distribution of W_9 is approximated by normal distribution. Let $W_9 = Z_9 - 0.4$. We have that $\mathbf{E}W_9 = 0$, $\operatorname{Var} W_9 = 0.25$,

$$\begin{aligned}
\mathbf{P}(Z_9 > 0.25) &= \mathbf{P}(W_9 + 0.4 > 0.45) = \mathbf{P}(W_9 > 0.05) \\
&= \mathbf{P}\left(\frac{W_9}{\sqrt{0.25}} > \frac{1}{\sqrt{0.25}}0.05\right) \sim \mathbf{P}\left(Z > \frac{1}{\sqrt{0.25}}0.05\right) \\
&= \mathbf{P}(Z \geq 0.05/0.5) \\
&= \mathbf{P}(Z \geq 0.01) = 1 - 0.5398 = 0.4602,
\end{aligned}$$

where $Z \sim N(0,1)$.

Solution of Problem F.7. The m.g.f. is

$$\begin{aligned}
M_X(t) = c\int_1^{\infty} 2^{-x}e^{tx}dx = c\int_1^{\infty} e^{tx - x\ln 2}dx &= c\frac{e^{tx-x\ln 2}}{t - \ln 2}\Big|_1^{\infty} \\
&= -c\frac{e^{t-\ln 2}}{t - \ln 2} = c\frac{e^{t-\ln 2}}{\ln 2 - t}.
\end{aligned}$$

This holds for $t < \ln 2$. From $M_X(0) = 1$, we obtain that

$$c\frac{e^{-\ln 2}}{\ln 2} = 1, \qquad c = \frac{\ln 2}{e^{-\ln 2}} = 2\ln 2 = 1.3863.$$

Hence

$$M_X(t) = c\frac{e^{t-\ln 2}}{\ln 2 - t} = \frac{e^t \ln 2}{\ln 2 - t}.$$

Hence $X = Y+1$, where Y has exponential distribution, i.e., $Y \sim Exp(\ln 2)$. Hence $\mathbf{E}(X) = 1 + 1/\ln 2 = 2.4427$ (in addition, $\operatorname{Var} X = 1/(\ln 2)^2 = 2.0814$).

Solution of Problem F.8. It can be noticed that

$$f(x) = (g * h)(x) = c\int_0^{\infty} ye^{-y}\mathbb{I}_{[-1,0]}(x-y)dy, \quad x > -1,$$

i.e., it is the convolution of

$$g(x) = cxe^{-x}\mathbb{I}_{[0,+\infty)}(x), \quad h(x) = \mathbb{I}_{[-1,0]}(x)$$

(in fact, $c = 1$ but we do not need this). It follows that $X = Y + Z$, where X and Y are independent, $Y \sim \Gamma(2,1)$ and $Z \sim U(-1,0)$. From the tables, we obtain that $\mathbf{E}Z = -1/2$; from the tables, we obtain that $\mathbf{E}Y = 2$. Hence $\mathbf{E}X = \mathbf{E}Y + \mathbf{E}Z = 2 - 1/2 = 1.5$ and $\operatorname{Var} X = \operatorname{Var} Y + \operatorname{Var} Z = 2 + 1/12$.

Solution of Problem F.9. We have that

$$\mathbf{E}(X+Y) = \mathbf{E}(X) + \mathbf{E}(Y).$$

$Z = X + Y \sim N(6,16)$. Since X and Y are independent random variables, we have that

$$\operatorname{Var}(X+Y) = \operatorname{Var}(X) + \operatorname{Var}(Y).$$

$Z = X + Y \sim N(6,16)$. Hence

$$\mathbf{P}(Z < 6.4) = \mathbf{P}\left(N(0,1) < \frac{6.4-6}{4}\right) = \mathbf{P}(N(0,1) < 0.1) = 0.5398.$$

Solution of Problem F.10. We have that $X_i \sim Exp(\lambda)$ with $\lambda = 2$. The c.d.f. for X_i is $F(x) = (1 - e^{-\lambda}x)\mathbb{I}_{[0,+\infty)}(x)$. It can be calculated as $\mathbf{P}(X_i > x) = \int_x^{\infty} \lambda e^{-\lambda s}ds = e^{-\lambda x}$. The p.d.f. for Y is

$$f_Y(y) = 3(1 - F(y))^2 f(y) = 3e^{-2\lambda y}\lambda e^{-\lambda y}\mathbb{I}_{[0,+\infty)}(y) = 3\lambda e^{-3\lambda y}\mathbb{I}_{[0,+\infty)}(y).$$

It is an exponential distribution again. Let $\alpha = 3\lambda$. The moment generating function is

$$M_Y(s) = \mathbf{E}e^{sY} = \alpha \int_0^\infty e^{sx}e^{-\alpha x}dx = \alpha \int_0^\infty e^{(s-\alpha)x}dx$$
$$= \frac{\alpha}{s-\alpha}e^{(s-\alpha)x}\Big|_0^\infty = -\frac{\alpha}{s-\alpha} = \frac{\alpha}{\alpha - s}.$$

It is defined for $s - \alpha < 0$, or $\alpha > s$. For $\lambda = 2$, it gives $\alpha = 6$,

$$M_Y(t) = \frac{6}{6-t}.$$

It is defined for $s < 6$.

Solution of Problem F.11. $\overline{S}$ is the sample variance. Hence the random variable $\frac{3}{1}S^2$ has χ_3^2-distribution (i.e. χ^2-distribution with three degrees of freedom), and

$$\mathbf{P}\left(S^2 \le 2.08\right) = \mathbf{P}\left(\frac{3}{1}S^2 \le 6.24\right) = \mathbf{P}\left(U \le 6.24\right) \sim 0.9,$$

where $U \sim \chi_3^2$.

Solution of Problem F.12. We have

$$\mathbf{E}(X) = 1, \quad \mathrm{Var}\, X = 1, \quad \mathbf{E}Z = 2, \quad \mathrm{Var}\, Z = 4.$$

We have

$$\mathbf{E}(X - \mathbf{E}X)(Z - \mathbf{E}Z) = \mathrm{corr}(X, Z)\sqrt{\mathrm{Var}\, X \mathrm{Var}\, Z} = -0.5\sqrt{2^2} = -1.$$

Hence

$$\mathbf{E}(XZ) - \mathbf{E}X\mathbf{E}Z = \mathbf{E}(X - \mathbf{E}X)(Z - \mathbf{E}Z) = -1.$$

Hence

$$\mathbf{E}(XZ) = \mathbf{E}X\mathbf{E}Z - 1 = 2 - 1 = 1.$$

Solution of Problem F.13. We have that $\frac{(n-1)}{\sigma^2}S^2 \sim \chi_{10}^2$. Let $\alpha = 0.1$. We have that $\chi_{\alpha/2}^2 = 18.307$ and $\chi_{1-\alpha/2}^2 = 3.940$.

$$\mathbf{P}\left(\chi_{1-\alpha/2}^2 < \frac{(n-1)}{\sigma^2}S^2 < \chi_{\alpha/2}^2\right) = 1 - \alpha.$$

Hence

$$\mathbf{P}\left(1/\chi_{\alpha/2}^2 < \frac{\sigma^2}{(n-1)S^2} < 1/\chi_{1-\alpha/2}^2\right) = 1 - \alpha,$$

$$\mathbf{P}\left(\frac{(n-1)S^2}{\chi^2_{\alpha/2}} < \sigma^2 < \frac{(n-1)S^2}{\chi^2_{1-\alpha/2}}\right) = 1-\alpha.$$

This gives

$$1-\alpha = \mathbf{P}\left(\frac{10S^2}{\chi^2_{\alpha/2}} < \sigma^2 < \frac{10S^2}{\chi^2_{1-\alpha/2}}\right) = \mathbf{P}\left(\frac{40}{\chi^2_{\alpha/2}} < \sigma^2 < \frac{40}{\chi^2_{1-\alpha/2}}\right)$$
$$= \mathbf{P}\left(\frac{40}{18.307} < \sigma^2 < \frac{40}{3.940}\right) = \mathbf{P}\left(2.1850 < \sigma^2 < 10.1523\right).$$

Therefore, the interval

$$[2.1850\,,\,10.1523]$$

is 90%-confidence interval for the true unknown σ^2.

Solution of Problem F.14. We have that $\frac{\overline{X}-\mu}{\sigma/\sqrt{n}} \sim N(0,1)$. Hence

$$\mathbf{P}(-z_{\alpha/2} \le \frac{\overline{X}-\mu}{\sigma/\sqrt{n}} \le z_{\alpha/2}) = 1-\alpha$$

$$\mathbf{P}(-\frac{z_{\alpha/2}\sigma}{\sqrt{n}} \le \overline{X}-\mu \le \frac{z_{\alpha/2}\sigma}{\sqrt{n}}) = 1-\alpha.$$

It can be rewritten as

$$\mathbf{P}(-\frac{z_{\alpha/2}\sigma}{\sqrt{n}} \le -\overline{X}+\mu \le \frac{z_{\alpha/2}\sigma}{\sqrt{n}}) = 1-\alpha.$$

It gives

$$\mathbf{P}(\overline{X}-\frac{z_{\alpha/2}\sigma}{\sqrt{n}} \le \mu \le \overline{X}+\frac{z_{\alpha/2}\sigma}{\sqrt{n}}) = 1-\alpha.$$

Therefore, the interval

$$\left[\overline{X}-\frac{z_{\alpha/2}\sigma}{\sqrt{n}},\, \overline{X}+\frac{z_{\alpha/2}\sigma}{\sqrt{n}}\right]$$

is $(1-\alpha)100\%$-confidence interval for the true unknown μ. Here $z_{\alpha/2}$ is $(1-\alpha/2)$-quantile for $N(0,1)$. We have that $z_{0.05} = 1.65$. Therefore, the interval

$$\left[3-\frac{1.65}{\sqrt{9}},\, 3+\frac{1.65}{\sqrt{9}}\right] = [2.45, 3.55]$$

is 90%-confidence interval for the true unknown μ.

Appendix 3: Some Bonus Challenging Problems

The problems below could be more challenging than listed above problems. The solutions are left to the reader to ponder; it may require some additional reading.

Problem B.1 *Assume that events A and B are independent, events B and C are independent, and events A and C are independent. Is it possible that the events A, B, C are not mutually independent?*

Problem B.2 *Is it possible to find an example of a probability distribution on $\mathbf{R}$ such that its support coincides with the set of all rational numbers? If yes, give an example, if no, prove it.*

Problem B.3 *Let $\mathbf{Q}^2$ be the set of all pairs $(x, y) \in \mathbf{R}^2$ such that both x and y are rational numbers. We consider a random direct line L in $\mathbf{R}^2$ such that $(0,0) \in L$ with probability 1, and that the angle between L and the vector $(1,0)$ has the uniform distribution on $[0, \pi)$. Find the probability that the set $L \cap \mathbf{Q}^2$ is finite.*

Problem B.4 *We used the following rule for the differentiation of an integral:*

$$\frac{d}{dz}\int_a^b f(x,z)dx = \int_a^b \frac{df}{dz}(x,z)dx.$$

It holds under certain assumptions for f. Find an example where this rule does not hold.

Problem B.5 *Find an example of discrete random variable such that $\sum_x |x|p(x) = \infty$, where $p(x)$ is the probability frequency function. (Hint: look for a random variable with infinite set of possible values.)*

Problem B.6 *Find an example of a discrete distribution such that the expectation is not defined (even cannot be regarded as $+\infty$ or $-\infty$).*

Problem B.7 *Let ξ_k be random variables such that $\mathbf{E}|\xi_k|^n < +\infty$, $k = 1, ..., n$. Prove that $\mathbf{E}|\xi_1 \xi_2 \cdots \xi_n| < +\infty$. Hint: use the so-called Hölder inequality (this inequality was not discussed in this course).*

Problem B.8 *Let a vector (X, Y) be uniformly distributed in the circle $D = \{(x, y): \ x^2 + y^2 \leq 1\} \subset \mathbf{R}^2$. Find the marginal density for X.*

Problem B.9 *Find an example of random variables $X \sim N(0,1)$, $Y \sim N(0,1)$ such that (X, Y) is not a Gaussian vector.*

Problem B.10 *Assume that X is a continuous random variable with density, and that Y is a discrete random variable. Prove that $X + Y$ is a continuous random variable with density.*

Problem B.11 *Let X be a random variable with the density*

$$f(x) = \begin{cases} c \int_{\max(0,x-1)}^{x} y e^{-y} dy, & x \geq 0 \\ 0, & x < 0, \end{cases}$$

where $c > 0$ is a constant. Calculate $\mathbf{E}(X)$.

Problem B.12 *Let $p = \max(p_1, p_2)$, where p_i are independent random variables uniformly distributed in $[0, 1]$. Let the distribution of X can be described as a so-called compound distribution such that X is a Bernoulli random variable conditionally given p such that*

$$X = \begin{cases} 1, & \text{with probability } p \\ 0, & \text{with probability 1-}p. \end{cases}$$

Find the unconditional distribution of X.

Problem B.13 *Let $\lambda \sim Exp(\theta)$ be a random variable, where $\theta > 0$ is a given real number. Let the distribution of X can be described as a so-called compound distribution such that $X \sim Exp(\lambda)$ conditionally given λ. Find the unconditional distribution of X.*

For the following two problems, you may use that

$$f_{\Theta|X}(\theta|x) \propto f_{X|\Theta}(x|\theta) \times f_{\Theta}(\theta),$$

for the conditional densities of two random vectors X and Θ and for the unconditional density $f_{\Theta}(\theta)$.

Problem B.14 *Consider random i.i.d. sample* $X = (X_1, ..., X_n)$ *such that* $X \sim G(a, \Theta)$ *given* Θ, *and prior distribution of the parameter is* $\Theta \sim Exp(\lambda)$. *Find conditional density for* Θ *given* X, *i.e., find* $f_{\Theta|X}(\theta|x)$ *(identify this distribution, if possible). Find* $\mathbf{E}(\Theta|X)$.

Problem B.15 *Let* X *be a random variable with conditionally exponential distribution* $Exp(\Lambda)$ *given* Λ. *Let* $\Lambda \sim U(0,1)$. *Find the conditional density* $f_{\Lambda|X}(\lambda|x)$ *for* λ *given* X. *Verify that* $\int_0^1 f_{\Lambda|X}(\lambda|x)d\lambda = 1$. *Find the conditional expectation* $\mathbf{E}(\Lambda|X = 1)$.

Appendix 4: Statistical Tables

For convenience, we provide below the tables for some distributions. These tables are sufficient for the examples and problems in this book.

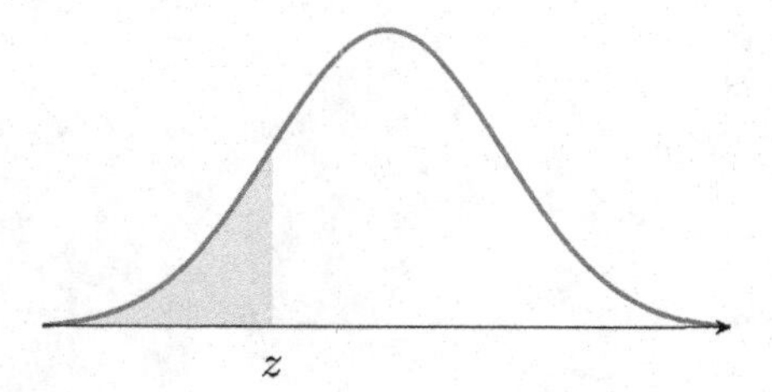

z	0	0.01	0.02	0.03	0.04	0.05	0.06	0.07	0.08	0.09
0	0.5	0.496	0.492	0.488	0.484	0.4801	0.4761	0.4721	0.4681	0.4641
-0.1	0.4602	0.4562	0.4522	0.4483	0.4443	0.4404	0.4364	0.4325	0.4286	0.4247
-0.2	0.4207	0.4168	0.4129	0.409	0.4052	0.4013	0.3974	0.3936	0.3897	0.3859
-0.3	0.3821	0.3783	0.3745	0.3707	0.3669	0.3632	0.3594	0.3557	0.352	0.3483
-0.4	0.3446	0.3409	0.3372	0.3336	0.33	0.3264	0.3228	0.3192	0.3156	0.3121
-0.5	0.3085	0.305	0.3015	0.2981	0.2946	0.2912	0.2877	0.2843	0.281	0.2776
-0.6	0.2743	0.2709	0.2676	0.2643	0.2611	0.2578	0.2546	0.2514	0.2483	0.2451
-0.7	0.242	0.2389	0.2358	0.2327	0.2296	0.2266	0.2236	0.2206	0.2177	0.2148
-0.8	0.2119	0.209	0.2061	0.2033	0.2005	0.1977	0.1949	0.1922	0.1894	0.1867
-0.9	0.1841	0.1814	0.1788	0.1762	0.1736	0.1711	0.1685	0.166	0.1635	0.1611
-1	0.1587	0.1562	0.1539	0.1515	0.1492	0.1469	0.1446	0.1423	0.1401	0.1379
-1.1	0.1357	0.1335	0.1314	0.1292	0.1271	0.1251	0.1230	0.1210	0.1190	0.1170
-1.2	0.1151	0.1131	0.1112	0.1093	0.1075	0.1056	0.1038	0.102	0.1003	0.0985
-1.3	0.0968	0.0951	0.0934	0.0918	0.0901	0.0885	0.0869	0.0853	0.0838	0.0823
-1.4	0.0808	0.0793	0.0778	0.0764	0.0749	0.0735	0.0721	0.0708	0.0694	0.0681
-1.5	0.0668	0.0655	0.0643	0.063	0.0618	0.0606	0.0594	0.0582	0.0571	0.0559
-1.6	0.0548	0.0537	0.0526	0.0516	0.0505	0.0495	0.0485	0.0475	0.0465	0.0455
-1.7	0.0446	0.0436	0.0427	0.0418	0.0409	0.0401	0.0392	0.0384	0.0375	0.0367
-1.8	0.0359	0.0351	0.0344	0.0336	0.0329	0.0322	0.0314	0.0307	0.0301	0.0294
-1.9	0.0287	0.0281	0.0274	0.0268	0.0262	0.0256	0.025	0.0244	0.0239	0.0233
-2	0.0228	0.0222	0.0217	0.0212	0.0207	0.0202	0.0197	0.0192	0.0188	0.0183
-2.1	0.0179	0.0174	0.017	0.0166	0.0162	0.0158	0.0154	0.015	0.0146	0.0143
-2.2	0.0139	0.0136	0.0132	0.0129	0.0125	0.0122	0.0119	0.0116	0.0113	0.011
-2.3	0.0107	0.0104	0.0102	0.0099	0.0096	0.0094	0.0091	0.0089	0.0087	0.0084
-2.4	0.0082	0.008	0.0078	0.0075	0.0073	0.0071	0.0069	0.0068	0.0066	0.0064
-2.5	0.0062	0.006	0.0059	0.0057	0.0055	0.0054	0.0052	0.0051	0.0049	0.0048
-2.6	0.0047	0.0045	0.0044	0.0043	0.0041	0.004	0.0039	0.0038	0.0037	0.0036
-2.7	0.0035	0.0034	0.0033	0.0032	0.0031	0.003	0.0029	0.0028	0.0027	0.0026
-2.8	0.0026	0.0025	0.0024	0.0023	0.0023	0.0022	0.0021	0.0021	0.002	0.0019
-2.9	0.0019	0.0018	0.0018	0.0017	0.0016	0.0016	0.0015	0.0015	0.0014	0.0014
-3	0.0013	0.0013	0.0013	0.0012	0.0012	0.0011	0.0011	0.0011	0.001	0.001

Table A1: Cumulative Standard Normal Distribution Table

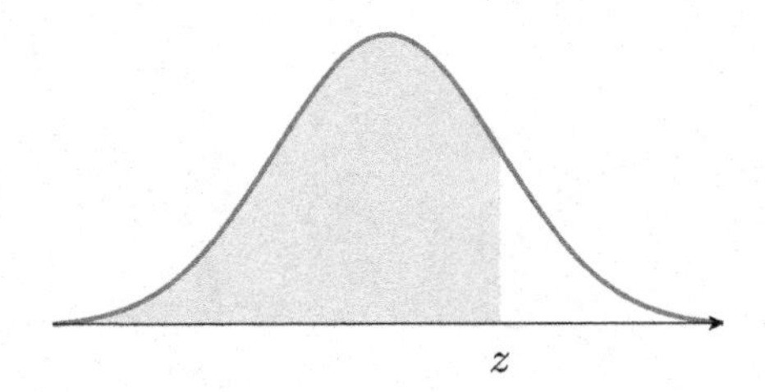

z	0	0.01	0.02	0.03	0.04	0.05	0.06	0.07	0.08	0.09
0	0.5000	0.5040	0.5080	0.5120	0.5160	0.5199	0.5239	0.5279	0.5319	0.5359
0.1	0.5398	0.5438	0.5478	0.5517	0.5557	0.5596	0.5636	0.5675	0.5714	0.5753
0.2	0.5793	0.5832	0.5871	0.590	0.5948	0.5987	0.6026	0.6064	0.6103	0.6141
0.3	0.6179	0.6217	0.6255	0.6293	0.6331	0.6368	0.6406	0.6443	0.6480	0.6517
0.4	0.6554	0.6591	0.6628	0.6664	0.6700	0.6736	0.6772	0.6808	0.6844	0.6879
0.5	0.6915	0.6950	0.6985	0.7019	0.7054	0.7088	0.7123	0.7157	0.7190	0.7224
0.6	0.7257	0.7291	0.7324	0.7357	0.7389	0.7422	0.7454	0.7486	0.7517	0.7549
0.7	0.7580	0.7611	0.7642	0.7673	0.7704	0.7734	0.7764	0.7794	0.7823	0.7852
0.8	0.7881	0.791	0.7939	0.7967	0.7995	0.8023	0.8051	0.8078	0.8106	0.8133
0.9	0.8159	0.8186	0.8212	0.8238	0.8264	0.8289	0.8315	0.8340	0.8365	0.8389
1	0.8413	0.8438	0.8461	0.8485	0.8508	0.8531	0.8554	0.8577	0.8599	0.8621
1.1	0.8643	0.8665	0.8686	0.8708	0.8729	0.8749	0.877	0.8790	0.8810	0.8830
1.2	0.8849	0.8869	0.8888	0.8907	0.8925	0.8944	0.8962	0.8980	0.8997	0.9015
1.3	0.9032	0.9049	0.9066	0.9082	0.9099	0.9115	0.9131	0.9147	0.9162	0.9177
1.4	0.9192	0.9207	0.9222	0.9236	0.9251	0.9265	0.9279	0.9292	0.9306	0.9319
1.5	0.9332	0.9345	0.9357	0.937	0.9382	0.9394	0.9406	0.9418	0.9429	0.9441
1.6	0.9452	0.9463	0.9474	0.9484	0.9495	0.9505	0.9515	0.9525	0.9535	0.9545
1.7	0.9554	0.9564	0.9573	0.9582	0.9591	0.9599	0.9608	0.9616	0.9625	0.9633
1.8	0.9641	0.9649	0.9656	0.9664	0.9671	0.9678	0.9686	0.9693	0.9699	0.9706
1.9	0.9713	0.9719	0.9726	0.9732	0.9738	0.9744	0.975	0.9756	0.9761	0.9767
2	0.9772	0.9778	0.9783	0.9788	0.9793	0.9798	0.9803	0.9808	0.9812	0.9817
2.1	0.9821	0.9826	0.983	0.9834	0.9838	0.9842	0.9846	0.985	0.9854	0.9857
2.2	0.9861	0.9864	0.9868	0.9871	0.9875	0.9878	0.9881	0.9884	0.9887	0.989
2.3	0.9893	0.9896	0.9898	0.9901	0.9904	0.9906	0.9909	0.9911	0.9913	0.9916
2.4	0.9918	0.992	0.9922	0.9925	0.9927	0.9929	0.9931	0.9932	0.9934	0.9936
2.5	0.9938	0.994	0.9941	0.9943	0.9945	0.9946	0.9948	0.9949	0.9951	0.9952
2.6	0.9953	0.9955	0.9956	0.9957	0.9959	0.996	0.9961	0.9962	0.9963	0.9964
2.7	0.9965	0.9966	0.9967	0.9968	0.9969	0.997	0.9971	0.9972	0.9973	0.9974
2.8	0.9974	0.9975	0.9976	0.9977	0.9977	0.9978	0.9979	0.9979	0.998	0.9981
2.9	0.9981	0.9982	0.9982	0.9983	0.9984	0.9984	0.9985	0.9985	0.9986	0.9986
3	0.9987	0.9987	0.9987	0.9988	0.9988	0.9989	0.9989	0.9989	0.999	0.999

Table A2: Cumulative Standard Normal Distribution Table (cont.)

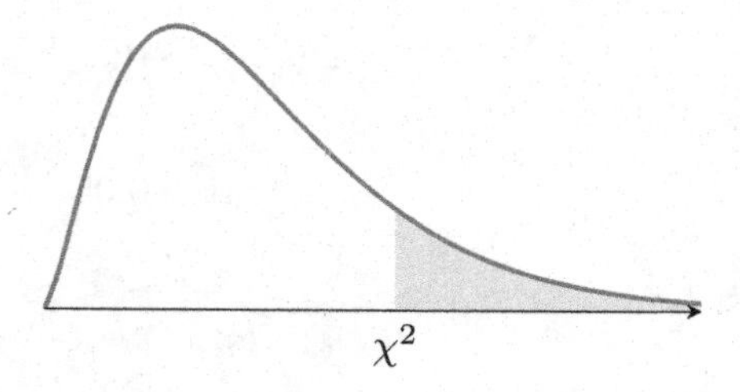

	0.995	0.99	0.975	0.95	0.9	0.1	0.05	0.025	0.01	0.005
$df = 1$	0.000	0.000	0.001	0.004	0.016	2.706	3.841	5.024	6.635	7.879
2	0.010	0.020	0.051	0.103	0.211	4.605	5.991	7.378	9.210	10.597
3	0.072	0.115	0.216	0.352	0.584	6.251	7.815	9.348	11.345	12.838
4	0.207	0.297	0.484	0.711	1.064	7.779	9.488	11.143	13.277	14.86
5	0.412	0.554	0.831	1.145	1.61	9.236	11.07	12.833	15.086	16.75
6	0.676	0.872	1.237	1.635	2.204	10.645	12.592	14.449	16.812	18.548
7	0.989	1.239	1.69	2.167	2.833	12.017	14.067	16.013	18.475	20.278
8	1.344	1.646	2.18	2.733	3.49	13.362	15.507	17.535	20.090	21.955
9	1.735	2.088	2.7	3.325	4.168	14.684	16.919	19.023	21.666	23.589
10	2.156	2.558	3.247	3.94	4.865	15.987	18.307	20.483	23.209	25.188
11	2.603	3.053	3.816	4.575	5.578	17.275	19.675	21.920	24.725	26.757
12	3.074	3.571	4.404	5.226	6.304	18.549	21.026	23.337	26.217	28.300
13	3.565	4.107	5.009	5.892	7.042	19.812	22.362	24.736	27.688	29.819
14	4.075	4.66	5.629	6.571	7.79	21.064	23.685	26.119	29.141	31.319
15	4.601	5.229	6.262	7.261	8.547	22.307	24.996	27.488	30.578	32.801
16	5.142	5.812	6.908	7.962	9.312	23.542	26.296	28.845	32.000	34.267
17	5.697	6.408	7.564	8.672	10.085	24.769	27.587	30.191	33.409	35.718
18	6.265	7.015	8.231	9.39	10.865	25.989	28.869	31.526	34.805	37.156
19	6.844	7.633	8.907	10.117	11.651	27.204	30.144	32.852	36.191	38.582
20	7.434	8.260	9.591	10.851	12.443	28.412	31.410	34.170	37.566	39.997
21	8.034	8.897	10.283	11.591	13.24	29.615	32.671	35.479	38.932	41.401
22	8.643	9.542	10.982	12.338	14.041	30.813	33.924	36.781	40.289	42.796
23	9.260	10.196	11.689	13.091	14.848	32.007	35.172	38.076	41.638	44.181
24	9.886	10.856	12.401	13.848	15.659	33.196	36.415	39.364	42.980	45.559
25	10.520	11.524	13.12	14.611	16.473	34.382	37.652	40.646	44.314	46.928
26	11.160	12.198	13.844	15.379	17.292	35.563	38.885	41.923	45.642	48.290
27	11.808	12.879	14.573	16.151	18.114	36.741	40.113	43.195	46.963	49.645
28	12.461	13.565	15.308	16.928	18.939	37.916	41.337	44.461	48.278	50.993
29	13.121	14.256	16.047	17.708	19.768	39.087	42.557	45.722	49.588	52.336
30	13.787	14.953	16.791	18.493	20.599	40.256	43.773	46.979	50.892	53.672

Table A3: Chi-Square Distribution Table

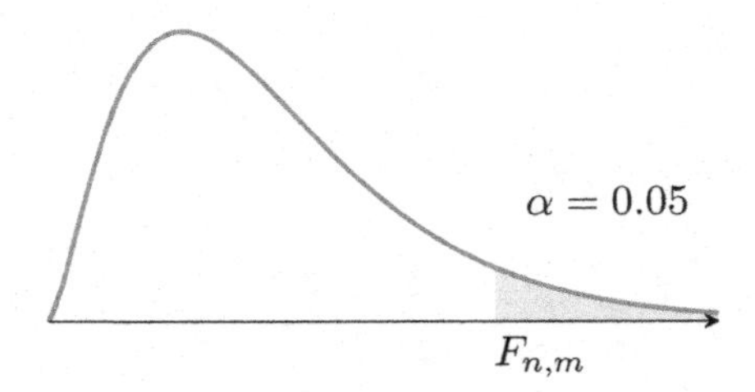

	$n=1$	2	3	4	5	6	7	8	9	10	12	15
$m=1$	161.5	199.5	215.7	224.6	230.2	234.0	236.8	238.9	240.5	241.9	243.9	246.0
2	18.51	19.00	19.16	19.25	19.30	19.33	19.35	19.37	19.39	19.40	19.41	19.43
3	10.13	9.552	9.277	9.117	9.014	8.941	8.887	8.845	8.812	8.786	8.745	8.703
4	7.709	6.944	6.591	6.388	6.256	6.163	6.094	6.041	5.999	5.964	5.912	5.858
5	6.608	5.786	5.410	5.192	5.050	4.950	4.876	4.818	4.773	4.735	4.678	4.619
6	5.987	5.143	4.757	4.534	4.387	4.284	4.207	4.147	4.099	4.060	4.000	3.938
7	5.591	4.737	4.347	4.120	3.972	3.866	3.787	3.726	3.677	3.637	3.575	3.511
8	5.318	4.459	4.066	3.838	3.688	3.581	3.501	3.438	3.388	3.347	3.284	3.218
9	5.117	4.257	3.863	3.633	3.482	3.374	3.293	3.230	3.179	3.137	3.073	3.006
10	4.965	4.103	3.708	3.478	3.326	3.217	3.136	3.072	3.020	2.978	2.913	2.845
11	4.844	3.982	3.587	3.357	3.204	3.095	3.012	2.948	2.896	2.854	2.788	2.719
12	4.747	3.885	3.490	3.259	3.106	2.996	2.913	2.849	2.796	2.753	2.687	2.617
13	4.667	3.806	3.411	3.179	3.025	2.915	2.832	2.767	2.714	2.671	2.604	2.533
14	4.600	3.739	3.344	3.112	2.958	2.848	2.764	2.699	2.646	2.602	2.534	2.463
15	4.543	3.682	3.287	3.056	2.901	2.791	2.707	2.641	2.588	2.544	2.475	2.403
16	4.494	3.634	3.239	3.007	2.852	2.741	2.657	2.591	2.538	2.494	2.425	2.352
17	4.451	3.592	3.197	2.965	2.810	2.699	2.614	2.548	2.494	2.450	2.381	2.308
18	4.414	3.555	3.160	2.928	2.773	2.661	2.577	2.510	2.456	2.412	2.342	2.269
19	4.381	3.522	3.127	2.895	2.740	2.628	2.544	2.477	2.423	2.378	2.308	2.234
20	4.351	3.493	3.098	2.866	2.711	2.599	2.514	2.447	2.393	2.348	2.278	2.203
21	4.325	3.467	3.073	2.840	2.685	2.573	2.488	2.421	2.366	2.321	2.250	2.176
22	4.301	3.443	3.049	2.817	2.661	2.549	2.464	2.397	2.342	2.297	2.226	2.151
23	4.279	3.422	3.028	2.796	2.640	2.528	2.442	2.375	2.320	2.275	2.204	2.128
24	4.260	3.403	3.009	2.776	2.621	2.508	2.423	2.355	2.300	2.255	2.183	2.108
25	4.242	3.385	2.991	2.759	2.603	2.490	2.405	2.337	2.282	2.237	2.165	2.089
26	4.225	3.369	2.975	2.743	2.587	2.474	2.388	2.321	2.266	2.220	2.148	2.072
27	4.210	3.354	2.960	2.728	2.572	2.459	2.373	2.305	2.250	2.204	2.132	2.056
28	4.196	3.340	2.947	2.714	2.558	2.445	2.359	2.291	2.236	2.190	2.118	2.041
29	4.183	3.328	2.934	2.701	2.545	2.432	2.346	2.278	2.223	2.177	2.105	2.028
30	4.171	3.316	2.922	2.690	2.534	2.421	2.334	2.266	2.211	2.165	2.092	2.015

Table A4: F Distribution Table for $\alpha = 0.05$

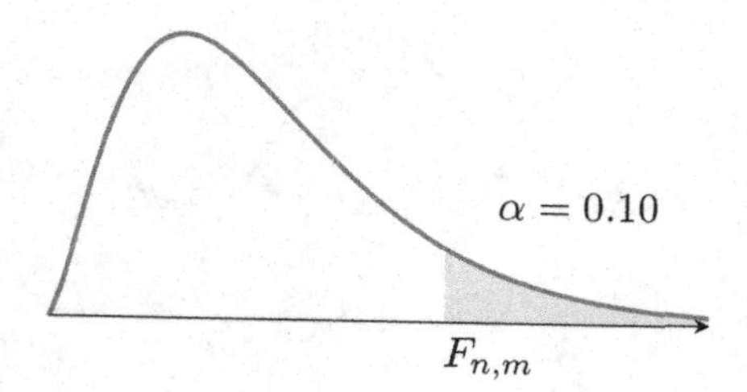

	$n=1$	2	3	4	5	6	7	8	9	10	12	15
$m=1$	39.86	49.50	53.59	55.83	57.24	58.2	58.91	59.44	59.86	60.19	60.71	61.22
2	8.526	9.000	9.162	9.243	9.293	9.326	9.349	9.367	9.381	9.392	9.408	9.425
3	5.538	5.462	5.391	5.343	5.309	5.285	5.266	5.252	5.240	5.230	5.216	5.200
4	4.545	4.325	4.191	4.107	4.051	4.010	3.979	3.955	3.936	3.920	3.896	3.870
5	4.060	3.780	3.619	3.520	3.453	3.405	3.368	3.339	3.316	3.297	3.268	3.238
6	3.776	3.463	3.289	3.181	3.108	3.055	3.014	2.983	2.958	2.937	2.905	2.871
7	3.589	3.257	3.074	2.961	2.883	2.827	2.785	2.752	2.725	2.703	2.668	2.632
8	3.458	3.113	2.924	2.806	2.726	2.668	2.624	2.589	2.561	2.538	2.502	2.464
9	3.360	3.006	2.813	2.693	2.611	2.551	2.505	2.469	2.440	2.416	2.379	2.340
10	3.285	2.924	2.728	2.605	2.522	2.461	2.414	2.377	2.347	2.323	2.284	2.244
11	3.225	2.860	2.660	2.536	2.451	2.389	2.342	2.304	2.274	2.248	2.209	2.167
12	3.177	2.807	2.606	2.480	2.394	2.331	2.283	2.245	2.214	2.188	2.147	2.105
13	3.136	2.763	2.560	2.434	2.347	2.283	2.234	2.195	2.164	2.138	2.097	2.053
14	3.102	2.726	2.522	2.395	2.307	2.243	2.193	2.154	2.122	2.095	2.054	2.010
15	3.073	2.695	2.490	2.361	2.273	2.208	2.158	2.119	2.086	2.059	2.017	1.972
16	3.048	2.668	2.462	2.333	2.244	2.178	2.128	2.088	2.055	2.028	1.985	1.940
17	3.026	2.645	2.437	2.308	2.218	2.152	2.102	2.061	2.028	2.001	1.958	1.912
18	3.007	2.624	2.416	2.286	2.196	2.130	2.079	2.038	2.005	1.977	1.933	1.887
19	2.990	2.606	2.397	2.266	2.176	2.109	2.058	2.017	1.984	1.956	1.912	1.865
20	2.975	2.589	2.380	2.249	2.158	2.091	2.040	1.999	1.965	1.937	1.892	1.845
21	2.961	2.575	2.365	2.233	2.142	2.075	2.023	1.982	1.948	1.920	1.875	1.827
22	2.949	2.561	2.351	2.219	2.128	2.061	2.008	1.967	1.933	1.904	1.859	1.811
23	2.937	2.549	2.339	2.207	2.115	2.047	1.995	1.953	1.919	1.890	1.845	1.796
24	2.927	2.538	2.327	2.195	2.103	2.035	1.983	1.941	1.906	1.877	1.832	1.783
25	2.918	2.528	2.317	2.184	2.092	2.024	1.971	1.929	1.895	1.866	1.820	1.771
26	2.909	2.519	2.307	2.174	2.082	2.014	1.961	1.919	1.884	1.855	1.809	1.760
27	2.901	2.511	2.299	2.165	2.073	2.005	1.952	1.909	1.874	1.845	1.799	1.749
28	2.894	2.503	2.291	2.157	2.064	1.996	1.943	1.900	1.865	1.836	1.790	1.740
29	2.887	2.495	2.283	2.149	2.057	1.988	1.935	1.892	1.857	1.827	1.781	1.731
30	2.881	2.489	2.276	2.142	2.049	1.980	1.927	1.884	1.849	1.819	1.773	1.722

Table A5: F Distribution Table for $\alpha = 0.10$

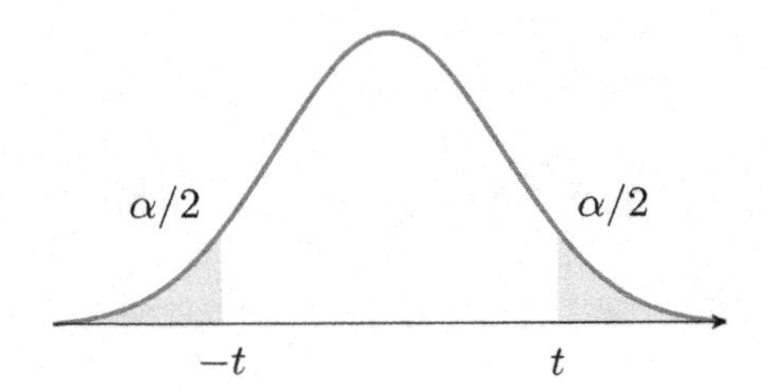

	$\alpha = 0.5$	0.4	0.3	0.2	0.1	0.05	0.02	0.01	0.002	0.001
$df = 1$	1.000	1.376	1.963	3.078	6.314	12.71	31.82	63.66	318.3	636.6
2	0.816	1.061	1.386	1.886	2.920	4.303	6.965	9.925	22.327	31.599
3	0.765	0.978	1.250	1.638	2.353	3.182	4.541	5.841	10.215	12.924
4	0.741	0.941	1.190	1.533	2.132	2.776	3.747	4.604	7.173	8.610
5	0.727	0.920	1.156	1.476	2.015	2.571	3.365	4.032	5.893	6.869
6	0.718	0.906	1.134	1.440	1.943	2.447	3.143	3.707	5.208	5.959
7	0.711	0.896	1.119	1.415	1.895	2.365	2.998	3.499	4.785	5.408
8	0.706	0.889	1.108	1.397	1.860	2.306	2.896	3.355	4.501	5.041
9	0.703	0.883	1.100	1.383	1.833	2.262	2.821	3.250	4.297	4.781
10	0.700	0.879	1.093	1.372	1.812	2.228	2.764	3.169	4.144	4.587
11	0.697	0.876	1.088	1.363	1.796	2.201	2.718	3.106	4.025	4.437
12	0.695	0.873	1.083	1.356	1.782	2.179	2.681	3.055	3.930	4.318
13	0.694	0.870	1.079	1.350	1.771	2.160	2.650	3.012	3.852	4.221
14	0.692	0.868	1.076	1.345	1.761	2.145	2.624	2.977	3.787	4.140
15	0.691	0.866	1.074	1.341	1.753	2.131	2.602	2.947	3.733	4.073
16	0.690	0.865	1.071	1.337	1.746	2.120	2.583	2.921	3.686	4.015
17	0.689	0.863	1.069	1.333	1.740	2.110	2.567	2.898	3.646	3.965
18	0.688	0.862	1.067	1.330	1.734	2.101	2.552	2.878	3.610	3.922
19	0.688	0.861	1.066	1.328	1.729	2.093	2.539	2.861	3.579	3.883
20	0.687	0.860	1.064	1.325	1.725	2.086	2.528	2.845	3.552	3.850
21	0.686	0.859	1.063	1.323	1.721	2.080	2.518	2.831	3.527	3.819
22	0.686	0.858	1.061	1.321	1.717	2.074	2.508	2.819	3.505	3.792
23	0.685	0.858	1.060	1.319	1.714	2.069	2.500	2.807	3.485	3.768
24	0.685	0.857	1.059	1.318	1.711	2.064	2.492	2.797	3.467	3.745
25	0.684	0.856	1.058	1.316	1.708	2.060	2.485	2.787	3.450	3.725
26	0.684	0.856	1.058	1.315	1.706	2.056	2.479	2.779	3.435	3.707
27	0.684	0.855	1.057	1.314	1.703	2.052	2.473	2.771	3.421	3.690
28	0.683	0.855	1.056	1.313	1.701	2.048	2.467	2.763	3.408	3.674
29	0.683	0.854	1.055	1.311	1.699	2.045	2.462	2.756	3.396	3.659
30	0.683	0.854	1.055	1.310	1.697	2.042	2.457	2.750	3.385	3.646

Table A6: Critical Values for Student's t Distribution

Bibliography

[1] Bradley, J.M., Tables and Formulae. Tables for use in tertiary institutions. Published by Murdoch University. 2005.

[2] Dokuchaev N.G. Mathematical finance: core theory, problems, and statistical algorithms. Routledge, 2007.

[3] Hoel, P.G., Port, S.C. & Stone, C.J. Introduction to Probability Theory, Houghton Mifflin, Boston 1971.

[4] Johnson, R.A. Bhattacharyya, G.K., Statistics: Principles and Methods, 7th Edition, Wiley, NJ, 2014.

[5] Rice, J.A. Mathematical Statistics and Data Analysis; 3rd edition. Duxbury, 2007.

[6] Shiryaev A. N. Boas, R.P. (Translator). Probability (Graduate Texts in Mathematics). Springer, 1995.

[7] Walpole, R.E., Myers, R.H., Myers, S.L, and Ye, K. Probability and Statistics for Engineers and Scientists; 9th edition. Prentice Hall, 2010.

Index

Legend of Notations and Abbreviations

- a.e. - almost everywhere, or for almost every
- a.s. - almost surely
- c.d.f. - cumulative distribution function
- $\mathrm{Cov}(X, Y)$ - covariance between X and Y
- $\mathrm{corr}\,(X, Y)$ - correlation between X and Y
- $\mathbf{E}X$ - expectation of X
- $\mathbf{E}X^2 = \mathbf{E}(X^2)$
- $\mathbf{E}XY = \mathbf{E}(XY)$
- $Exp(\lambda)$ - Exponential distribution with mean $\frac{1}{\lambda}$
- iff - if and only if
- i.i.d. - independent identically distributed
- $\mathbb{I}_A$ is the indicator function of an event A
- $\mathbb{I}_D(x)$ is the indicator function of the set D
- m.g.f. - moment generating function
- MLE - Maximum likelihood estimation
- $N(a, \sigma^2)$ - the normal distribution with mean a and variance σ^2
- p.d.f. - probability density function
- $U(a, b)$ - Uniform distribution with interval (a, b)
- $\mathrm{Var}\,X$ - variance of X
- $\emptyset$ - empty set
- $\{a, b\}$ - set with the elements a and b
- $[a, b]$ - set of all $x \in \mathbf{R}$ such that $a \le x \le b$
- 2^Ω - the set of all subsets of Ω
- $\mathbf{R}$ - the set of real numbers
- $\mathbf{R}^n$ - the set of real vectors (vector columns) with n components
- $x \triangleq X$ - means that x is defined such that $x = X$
- $\xi \sim N(0, v^2)$ - the random variable ξ has the distribution law $N(0, v^2)$
- $\propto$ - proportionality of p.d.f.
- $\forall$ - for all
- $\exists$ - exist
- χ_n^2 - Chi-square distribution with n degrees of freedom ($df = n$)

Printed in the United States
By Bookmasters